Marco Chiodi

An Innovative 3D-CFD-Approach towards
Virtual Development of Internal Combustion Engines

VIEWEG+TEUBNER RESEARCH

Marco Chiodi

An Innovative 3D-CFD-Approach towards Virtual Development of Internal Combustion Engines

VIEWEG+TEUBNER RESEARCH

Bibliographic information published by the Deutsche Nationalbibliothek
The Deutsche Nationalbibliothek lists this publication in the Deutsche Nationalbibliografie;
detailed bibliographic data are available in the Internet at http://dnb.d-nb.de.

Dissertation der Universität Stuttgart, Fakultät Konstruktions-,
Produktions-und Fahrzeugtechnik, 2010

Hauptberichter: Prof. Dr.-Ing. Michael Bargende
Mitberichter: Prof. Dr.-Ing. Giorgio Rizzoni
Mitberichter: Prof. Dr.-Ing. Federico Millo
Tag der mündlichen Prüfung: 22. Juni 2010

1st Edition 2011

Editorial Office: Stefanie Brich | Sabine Schöller

Vieweg+Teubner Verlag is a brand of Springer Fachmedien.
Springer Fachmedien is part of Springer Science+Business Media.
www.viewegteubner.de

Cover design: KünkelLopka Medienentwicklung, Heidelberg
Printing company: STRAUSS GMBH, Mörlenbach
Printed on acid-free paper
Printed in Germany

ISBN 978-3-8348-1540-8

"O frati", dissi "che per cento milia
perigli siete giunti a l'occidente,
a questa tanto picciola vigilia
d'i nostri sensi ch'è del rimanente,
non vogliate negar l'esperienza,
di retro al sol, del mondo sanza gente.
Considerate la vostra semenza:
fatti non foste a viver come bruti,
ma per seguir virtute e canoscenza".

Dante Alighieri
La Divina Commedia,
Inferno XXVI (Ulisse, 112-120)

Acknowledgments

This work has been carried out during my activity in the department of internal combustion engines of the FKFS and IVK-University of Stuttgart under the direction of Prof. Dr.-Ing. Michael Bargende. Since 1998 first as research assistant and then as project manager of 3D-CFD-simulations I have dedicated most of my time in the development of a new approach in the 3D-CFD-simulation of internal combustion engines, that aims a better integration in the engine development process. This work wants to be the report of this project.

I specially want to thank Prof. Michael Bargende for the opportunity, motivation, scientific support and often also scientific inspiration he gave me during all the years. Many thanks to Prof. Giorgio Rizzoni from the Ohio State University (USA) and Prof. Federico Millo from the Politecnico di Torino (Italy) first of all for having willingly accepted to be the co-referent of my work and then for their reliable support and professional opinions all the time.

Also a special thank goes to the Head of Engine Development at Volkswagen Motorsport: Dr.-Ing. Donatus Wichelhaus for his extraordinary innovative thinking, talent and openness that have permitted to successfully realize many interesting projects together. I want to thank him also for having allowed the publication of some results in this work.

Thanks to the colleagues of my team: Oliver Mack (nickname: Meshelangelo, because he is the Michelangelo of the meshes) and Alessandro Ferrari for the wonderful collaboration. Thanks also to the other colleagues at the FKFS and IVK-University of Stuttgart (in particular Hans-Jürgen Berner, Gerd Hitzler, Dr. Dietmar Schmidt and Dr. Michael Grill) for many interesting scientific discussions and their suggests during all the years that we have worked together.

Stuttgart, May 2010

Marco Chiodi

Table of Contents

Abstract

Society is a dynamic being and transportation is a key factor for fulfilling the human needs. Among different means of transport, vehicles with internal combustion engines represent the most relevant share in global transportation and namely cars and motorcycles are a self-evident object in our daily activities. In the future the limitations on the environmental impact, especially concerning private traffic, will become harder and harder, i.e. this process towards a real sustainable mobility will dictate the future development steps in the automotive sector. Consequently the engine of the future will be light-weight, small, turbocharged, probably with an unconventional combustion strategy, silent, with extremely low exhaust emissions (also at cold start), very low fuel consumption and CO_2 emission, high specific power, "vigorous" torque at low engine speed, long-life reliability and, maybe the most difficult target, it has to be still affordable. In order to meet the expectations of these complex tasks, engine simulation will play a key role in the future engine-development-process.

During the years engine simulation has continuously gained in reliance but mainly in the last two decades the rapid increase of computer performance has boosted the utilizability of simulation programs. At the present time the application of simulation programs towards a reliable "virtual engine development" represents one of the greatest challenges. An all-embracing tool for a reliable, detailed and predictive simulation of an internal combustion engine does not exist and will probably never exist. But there is another more realistic way represented by a selective coupling of different simulation tools depending on the different targeted tasks of the engine development. For the performance of this coupling the 3D-CFD-approach (in particular related to an analysis of the engine operating cycle) is a decisive component.

The 3D-CFD-simulation represents undeniably the most detailed approach for the investigation of the engine operating cycle. From its basic concept this approach should allow an unlimited predictability of engine processes by unrestricted varying of any parameter of both the engine setting and the operating condition. Depending on the chosen degree of mesh discretization it is possible to fully record the fluid motion and any chemical and thermodynamic phenomena acting in any part of the 3D-CFD-domain up to a length scale which is not far away from molecular dimensions. But this is the theory, the practice looks quite different. The computational resources in terms of the required hardware and the related CPU-time are very often prohibitive. In addition the setting of initial and boundary conditions can easily introduce sources of inaccuracy that irremediably compromises the overall quality of the results and drastically reduces the benefits of such resource investments. Another critical point is represented by few three-dimensional engine process models that due to the lack of phenomena understanding at the fundamental physical level, inaccurate mathematical formulations, numerical dependencies on the mesh structure, ambiguous validation processes, etc., are not able to ensure a high level of reliability in reproducing and predicting the requested engine processes.

In this work the development of a new 3D-CFD-tool called *QuickSim* as a new approach in the three dimensional analysis of internal combustion engines is presented. This tool tries to take advantage of the potentiality of the traditional 3D-CFD-approach combined with a remarkable reduction of the above mentioned drawbacks so that a major contribution in engine development process can be ensured. The peculiarities of *QuickSim* are: fast analysis calculations, reliability, user-friendliness, clear representation of the results without ambiguity and cost efficiency. Moreover this simulation tool aims for a higher integration into the existent engine development process so that a more efficient comparison with both experimental data and other simulation programs can be achieved.

More precisely *QuickSim* is a 3D-CFD-tool exclusively dedicated and optimized for the simulation of internal combustion engines (gasoline, diesel, CNG and other alternative fuels). There are no limitations regarding fuel injection and valve motion strategies. This fast response simulation, thanks to improved or newly developed 3D-CFD-models for the description of engine processes, ensures an efficient and reliable calculation also using coarse 3D-CFD-meshes. Based on this approach the CPU-time can be reduced up to a factor 100 in comparison to traditional 3D-CFD-simulations. An integrated and automatic "evaluation tool" establishes a comprehensive analysis of the relevant engine parameters (clear representation of the results) and the "internal coupling" with the real working-process analysis (WP) allows both a supported analysis of the engine processes and a better comparison and control with test bench results and other simulation tools. Furthermore the simulation of several successive operating cycles (reduction up to completely elimination of the influence of the initial conditions) and the extension of the simulated 3D-CFD-domain up to the full engine (increasing of predictability and reduction of the influence of boundary conditions) are possible.

As introduced before, the 3D-CFD-models implemented into *QuickSim* do not have a general validity for any thermodynamic investigation. Their formulation is adjusted in order to both optimize the solution of engine processes and reduce the computational resources required for the calculation. These models are based on a combination of different approaches (traditional local 3D-CFD-models, engine-specific phenomenological relationships, trained neural networks, databases and if needed empirical relationships) and take explicitly into account the cell dimensions and cell structures. This combination allows a reliable analysis of the process from its relevant behavior for practical applications in the engine design process; namely the thermodynamic aspect. The "internal coupling" between the 3D-CFD-calculation and the real-working-process analysis (WP) supports the local modeling and allows to compare the "global" results (heat release, wall heat transfer, internal energy variations, etc.) at each time-step so that any implausible differences can be immediately recognized and eventually corrected.

This work explains the basic idea and the modeling in *QuickSim*. In particular the focus is on the modeling of the thermodynamic properties of the working fluid, the combustion process and the wall heat-transfer.

The thermodynamic properties of the working fluid are determined using a combination of few scalars (six numerical species) for the description of the mixture composition for any arbitrary fuel $C_nH_mO_rN_q$ and a "dynamic" calculation of the real gas constant, the enthalpy and the heat-release terms. As well known the thermodynamic properties of a combustion gas, first of all depend on its chemical composition, i.e. the result of complex reaction mechanisms (for common fuel more than thousand chemical reactions and about 7000 involved species). Since the solution of detailed combustion mechanisms would lead to exorbitant CPU-overheads, in *QuickSim* the species conservation equations are efficiently solved only for six scalars and the properties of the working fluid are loaded from databases or trained neural networks. E.g. the stored values in databases are the result of a reduced reaction mechanism based on chemical equilibriums that include also dissociation effects. These values are listed as functions of pressure, temperature and air/fuel ratio and can be easily addressed with fast algorithms. The preparation of these databases (unique for a given fuel composition and lower heating value), if not already existing, is performed at once during the pre-processing step within few hours. The aim of the fast calculation of the thermodynamic properties of the working fluid in *QuickSim* is to approach the properties of the real gas, as exact as possible, from the thermodynamic point of view, i.e. the calculated fluid composition that also may include pollutant species (NO_X and HC) cannot be properly used for the determination of exhaust emissions (this can be done only with the implementation of dedicated models for emission calculation).

In *QuickSim*, as in most of the 3D-CFD-calculations, the combustion model for SI-engines with partially premixed mixture is actually a heat-release model based on the prediction of the flame front propagation. Since the flame speed is remarkably influenced by the local mesh structure, dimension as well as orientation, a new approach in the implementation of an existing combustion model has been proposed towards a more reliable and less mesh-dependent calculation. Among other things, this approach takes into account the intrinsic incapability of the cells passed by the flame to provide reliable inputs to all models that explicitly require information about the unburned (e.g. the model for the calculation of the laminar flame speed) and burned zone at the flame front. In order to bridge this lack of capability in each cell of the 3D-CFD-mesh affected by the flame front, a continuously thermodynamic zero-dimensional splitting into a burned und unburned zone takes place so that the consistency of modeling can be increased.

In *QuickSim*, despite the typical approach used in 3D-CFD-calculations, the wall heat-transfer has been modeled using a local implementation of phenomenological approaches developed for the real working-process analysis. The calculation of the wall heat-transfer using this new 3D-CFD-model is continuously supported by the "internal coupling" between the 3D-CFD-simulation and the real working-process analysis that aims to ensure a very high predictability also in the 3D-CFD-simulation. Here it has been recognized that the wall heat-transfer, as the thermodynamic boundary condition of the combustion chamber, represents the main source of inaccuracy in the 3D-CFD-analysis during the working period of the engine operating cycle. Very often in the 3D-CFD-simulation attention is mainly paid to the modeling of phenomena like combustion (chemical kinetic reactions) and fuel spray formation, instead of ensuring a comparable level of accuracy in the solution of all relevant engine processes. This is a fatal error that easily compromises the quality of the simulation results because a simulation running with wrong thermodynamic boundary conditions will never deliver reliable results.

In the last chapter a recent application of *QuickSim* on a turbocharged Compressed Natural Gas race engine from Volkswagen Motorsport will be presented. The focus here is mainly on the analysis of different approaches in the 3D-CFD-simulation for supporting the engine development process: from a "simple" three-dimensional visualization and understanding of the occurring phenomena within the combustion chamber or the airbox up to a reliable virtual engine development. The discussion of these approaches aims to underline the advantage, drawbacks, reliability and predictability of each of them.

Zusammenfassung

Die Gesellschaft ist ein dynamisches System und Mobilität ist ein wesentlicher Faktor, um die Bedürfnisse der Menschen zu befriedigen. Unter den verschiedenen Transportmitteln, stellen Fahrzeuge mit Verbrennungsmotoren den größten Anteil am globalen Verkehrsaufkommen dar, und namentlich Autos und Motorräder sind ein selbstverständlicher Bestandteil unseres täglichen Lebens. In der Zukunft werden Umweltauflagen betreffend den Individualverkehr noch härter werden, d.h. die Entwicklung hin zu einer nachhaltigen Mobilität wird für die künftigen Fortschritte der Automobilbranche maßgeblich sein. Folglich wird der Motor der Zukunft folgende Eigenschaften besitzen: leicht, klein, aufgeladen, wahrscheinlich mit ungewöhnlicher Verbrennungsstrategie, leise, extrem abgasarm (auch beim Kaltstart), verbrauchs- und CO_2-günstig, hohe spezifische Leistung, starkes Drehmoment bei niedriger Leistung und zu guter Letzt bezahlbar. Um diese Herausforderungen bewältigen zu können, wird der Simulation eine Schlüsselrolle in der weiteren Motorenentwicklung zufallen.

In der Vergangenheit hat die Motorsimulation zunehmend an Bedeutung gewonnen, aber hauptsächlich in den letzten beiden Jahrzehnten verhalf die große Leistungssteigerung der Computer der Anwendung von Simulationsprogrammen zu einem enormen Aufschwung. Heutzutage stellt die Entwicklung der Simulationsprogramme hin zu einer „virtuellen Motorentwicklung" eine der größten Herausforderungen dar. Ein allumfassendes Werkzeug für eine zuverlässige, detaillierte und voraussagende Simulation eines Verbrennungsmotors existiert nicht und wird wahrscheinlich nie existieren. Aber es gibt einen weiteren, realistischeren Weg, der in Abhängigkeit des angestrebten Entwicklungsziels darin besteht, eine Kopplung ausgewählter Simulationsprogramme durchzuführen. Für die Leistungsfähigkeit dieser Kopplung stellt die 3D-CFD-Methode (insbesondere im Hinblick auf die Berechnung eines Motorzyklus) eine entscheidende Komponente dar.

Die 3D-CFD-Simulation stellt unbestreitbar den detailliertesten Zugang zur Untersuchung des Arbeitsspiels eines Motors dar. Prinzipiell sollte die Methode in der Lage sein, eine uneingeschränkte Vorhersagbarkeit der Motorprozesse bei beliebiger Variation der Parameter zu erlauben. In Abhängigkeit vom Grad der Netzfeinheit, ist es möglich sowohl die Bewegung des fluiden Mediums als auch jegliches chemisches oder thermodynamisches Phänomen in jedem Bereich des Strömungsgebiets aufzuzeichnen, und zwar bis zu einer Größenskala, die nicht weit oberhalb molekularer Dimensionen liegt. Soweit die Theorie, die Praxis sieht jedoch anders aus. Die Anforderungen an die benötigte Hardware und die damit verbundene Rechenzeit stellen sehr oft große Hindernisse dar. Zusätzlich kann das Einbinden der Anfangs- und Randbedingungen leicht zu Ungenauigkeiten führen, die die Gesamtqualität der Ergebnisse unumstößlich mindern und damit den Nutzen solcher Forschungsanstrengungen stark abschwächen. Ein weiterer kritischer Punkt liegt in einigen 3D-Modellen des Motors, die kein hohes Maß an Zuverlässigkeit bezüglich Reproduzierbarkeit und Vorhersagefähigkeit der entsprechenden Motorprozesse erlauben. Die Gründe dafür können u.a. Mängel in Bezug auf das Verständnis der physikalischen Grundlagen, ungenaue mathematische Formulierungen, numerische Netzabhängigkeiten oder mehrdeutige Validierungen sein.

In dieser Veröffentlichung soll nun die Entwicklung eines neuen 3D-CFD-Werkzeugs - *QuickSim* - vorgestellt werden, welches einen neuen Ansatz zur dreidimensionalen Analyse von Verbrennungsmotoren darstellt. *QuickSim* versucht das Potential des traditionellen 3D-CFD-Ansatzes bei gleichzeitiger Minimierung der erwähnten Nachteile auszunutzen, so dass ein bedeutender Beitrag zur Motorenentwicklung gesichert werden kann. Die Besonderheiten von *QuickSim* sind schnelle Berechnung, Verlässlichkeit, Benutzerfreundlichkeit, übersichtliche Ergebnisdarstellung ohne Mehrdeutigkeiten und Kosteneffizienz. Darüberhinaus ist es das Ziel von *QuickSim*, eine bessere Einbindung in die bestehende Motorentwicklung zu gewährleisten, so dass ein effektiverer Vergleich sowohl mit gemessenen als auch mit Daten anderer Simulationsprogramme erreicht wird.

Genauer gesagt stellt *QuickSim* ein 3D-CFD-Werkzeug dar, das ausschließlich der Simulation von Verbrennungsmotoren (Benzin, Diesel, CNG und andere alternative Kraftstoffe) gewidmet ist und dafür optimiert wurde. Es gibt keinerlei Einschränkungen bezüglich der Kraftstoffeinspritzung oder der Ventilhubverläufe. Diese schnelle Simulation garantiert dank verbesserter oder neu entwickelter 3D-CFD-Modelle für die Beschreibung der innermotorischen Prozesse eine effiziente und zuverlässige Berechnung, auch bei der Verwendung grober Netze. Basierend auf diesem Ansatz kann die benötigte CPU-Zeit im Vergleich mit einer traditionellen 3D-CFD-Rechnung bis zu einem Faktor 100 reduziert werden. Ein integriertes, automatisches Tool liefert eine verständliche Auswertung der relevanten Motorparameter (klare Darstellung der Ergebnisse), und die interne Kopplung mit einer Arbeitsprozessrechnung erlaubt sowohl eine Unterstützung der Analyse der Motorprozesse, als auch einen besseren Vergleich mit, bzw. eine bessere Kontrolle anhand von Prüfstandsmessungen oder anderen Simulationsprogrammen. Des Weiteren ist die Simulation mehrerer aufeinanderfolgender Arbeitsspiele (Reduzierung bzw. vollständige Eliminierung der Einflüsse der Anfangsbedingungen) und die Ausdehnung des untersuchten 3D-Strömungsgebietes auf den kompletten Motor (Erhöhung der Vorhersagefähigkeit und Reduzierung der Einflüsse der Randbedingungen) möglich.

Wie bereits erwähnt, besitzen die in *QuickSim* eingebundenen 3D-CFD-Modelle keine allgemeine Gültigkeit für beliebige thermodynamische Untersuchungen. Ihre Formulierung ist angepasst, um sowohl die Berechnung von Motorprozessen zu optimieren, als auch die benötigten Rechnerkapazitäten zu vermindern. Die Modelle basieren auf einer Kombination verschiedener Herangehensweisen (traditionelle 3D-CFD-Modelle, motorspezifische phänomenologische Zusammenhänge, neuronale Netze, Datenbanken und evtl. empirische Beziehungen) und berücksichtigen explizit die Zelldimensionen und -strukturen des Netzes. Diese Kombination erlaubt eine zuverlässige Analyse des für praktische Anwendungen relevanten Motorverhaltens in den Entwicklungsphase: namentlich ist dies der thermodynamische Aspekt. Die interne Kopplung von 3D-CFD- und Arbeitsprozessrechnung unterstützt die lokalen Modelle und erlaubt zu jedem Zeitpunkt einen Vergleich mit den „globalen" Ergebnissen (Wärmefreisetzung, Wandwärmeübergang, Änderungen der inneren Energie, usw.), so dass unplausible Unterschiede sofort erkannt und eventuell korrigiert werden können.

Diese Arbeit erklärt die Grundidee und die Modellierung von *QuickSim*. Der Fokus liegt hierbei vor allem auf den Modellen bezüglich des thermodynamischen Verhaltens, der Verbrennung und des Wandwärmeübergangs.

Die thermodynamischen Eigenschaften des Mediums werden durch eine Kombination von sechs sogenannten Skalaren (numerische Gattungen) zur Beschreibung der Gemischzusammensetzung eines beliebigen Kraftstoffs und einer dynamischen Berechnung der realen Gaskonstante, der Enthalpie und der Wärmefreisetzung bestimmt. Wie allgemein bekannt, hängen die thermodynamischen Eigenschaften eines Verbrennungsgases in erster Linie von seiner chemischen Zusammensetzung und den komplexen Reaktionsmechanismen (üblicherweise über 1000 unterschiedliche chemische Reaktionen und mehr als 7000 beteiligte Verbindungen) ab. Da die detaillierte Lösung der Verbrennungsmechanismen zu einer übermäßigen CPU-Belastung führen würde, werden die Erhaltungsgleichungen der chemischen Verbindungen in *QuickSim* nur für sechs Gattungen gelöst und die Eigenschaften des Arbeitsmediums stattdessen Datenbanken oder neuronalen Netzen entnommen. So sind z.B. die in Datenbanken gespeicherten Werte das Ergebnis eines reduzierten Reaktionsmechanismus, der auf chemischen Gleichungen basiert, welche auch Dissoziationseffekte berücksichtigen. Diese Werte existieren als Funktionen des Drucks, der Temperatur und des Luft-Kraftstoff-Verhältnisses und können mit Hilfe schneller Algorithmen auf einfache Weise übernommen werden. Falls diese Datenbanken (einzigartig für eine vorgegebene Kraftstoffzusammensetzung und einen bestimmten unteren Heizwert) nicht schon existieren, können sie im Rahmen der Simulationsvorbeitung innerhalb weniger Stunden erzeugt werden. Das Ziel dieser vereinfachten Ermittlung der thermodynamischen Eigenschaften des Mediums in *QuickSim* ist es, die Eigenschaften des realen Gases so gut wie möglich zu erfassen. Dies geschieht jedoch vor allem aus thermodynamischer Sicht, so dass die berechnete Zusammensetzung des Mediums, das auch Schadstoffe wie NO_X oder HC enthalten kann, nicht für eine korrekte Bestimmung der Abgasemissionen benutzt werden kann. Dies ist ausschließlich durch die Einbindung von Modellen möglich, die speziell der Berechnung der Emissionen gewidmet sind.

Wie in den meisten 3D-CFD-Simulationen, ist das Verbrennungsmodell für vorgemischte Otto-Motoren auch in *QuickSim* letztendlich ein Modell zur Wärmefreisetzung, das auf der Vorhersage der Flammenausbreitung basiert. Da die Flammengeschwindigkeit stark von der lokalen Netzstruktur (Zellgröße und -form) abhängt, wurde ein neuer Ansatz zur Einbindung eines bestehenden Verbrennungsmodells gemacht, der zu einer zuverlässigeren und weniger netzabhängigen Berechnung führt. Unter anderem berücksichtigt der Ansatz die inne-wohnende Unfähigkeit der Zellen, die von der Flamme erfasst werden, zuverlässige Eingangsdaten für diejenigen Modelle (z.B. das Modell zur Berechnung der laminaren Flammengeschwindigkeit) zu liefern, die explizite Informationen über die unverbrannte und verbrannte Zone im Bereich der Flammenfront benötigen. Um diese Lücke zu füllen, findet in jeder von der Flammenfront betroffenen Zelle eine kontinuierliche, thermodynamische, null-dimensionale Aufteilung in eine verbrannte und eine unverbrannte Zone statt, so dass die Richtigkeit der Modellierung erhöht werden kann.

Entgegen des in 3D-CFD-Rechnungen üblichen Ansatzes, wurde der Wandwärmeübergang in *QuickSim* mittels einer lokalen Anwendung phänomenologischer, für die Arbeitsprozess-rechnung entwickelter Ansätze eingebunden. Die Berechnung des Wandwärmeübergangs mit diesem neuen Modell wird durch die interne Kopplung von 3D-CFD- und Arbeitsprozess-rechnung fortwährend unterstützt, damit auch in der 3D-CFD-Simulation eine hohe Vorhersagbarkeit erzielt werden kann. Dabei wurde festgestellt, dass in der 3D-CFD-Analyse der Wandwärmeübergang als thermodynamische Randbedingung des Brennraums, die Hauptquelle für Ungenauigkeiten während der Hochdruckphase darstellt. Sehr oft wird in der 3D-CFD-Rechnung der Modellierung von Phänomenen wie Verbrennung oder Einspritzung große Aufmerksamkeit gewidmet, anstatt dafür Sorge zu tragen, dass die Lösung aller relevanten Motorprozesse eine vergleichbare Genauigkeit aufweist. Dies stellt einen unverzeihlichen Fehler dar, da die Simulationsergebnisse dadurch sehr leicht verfälscht werden: eine Simulation mit falschen thermodynamischen Randbedingungen wird niemals zuverlässige Ergebnisse liefern.

Im letzten Kapitel wird eine aktuelle *QuickSim*-Simulation eines turboaufgeladenen CNG-Rennmotors der Firma Volkswagen Motorsport vorgestellt. Der Augenmerk liegt dabei hauptsächlich in der Analyse verschiedener Ansätze der 3D-CFD-Simulation zur Unterstützung der Motorentwicklung: von einer einfachen dreidimensionalen Visualisierung, einem Verständnis für die Vorgänge im Brennraum oder der Airbox bis hin zu einer zuverlässigen, virtuellen Motorentwicklung. Die Auseinandersetzung mit diesen Ansätzen zielt darauf ab, ihre jeweiligen Vor- und Nachteile, ihre Zuverlässigkeit und ihre Vorhersagefähigkeit hervorzuheben.

Symbols, Subscripts and Abbreviations

Roman Symbols

a	-	Coefficient in the Wiebe combustion function
a	-	Constant in the Re-Nu correlation
$\bar{a}$	-	Constant in the phenomenological 3D-CFD-heat-transfer
A	m^2	Area
A	-	Pre-exponential factor of the Arrhenius-equation
$[A]$	-	Sub-term in the combustion term of Bargende's heat transfer formulation
$[A]$	mol/m^3	Species concentration in a reaction scheme
A_{comb}	-	Corrector factor for the turbulence velocity in the modeling of the flame propagation model (combustion model)
A_f	m^2	Area of the flame front
$A_{f,L}$	m^2	Area of the laminar wrinkled flame front (flamelet model)
$A_{f,T}$	m^2	Area of the equivalent turbulent flame front
a_j	-	Tabulated coefficients for the numerical formulation of fluid thermo-physical properties
A_j	m^2	Area of a face of a discrete element (cell) j
A_{Geom}	m^2	Geometric flow area of injector orifice
A_W	m^2	Area of the combustion chamber
B	m	Cylinder bore
$[B]$	-	Sub-term in the combustion term of Bargende's heat transfer formulation
c	-	Progress variable
C	-	Constant in Hohenberg's heat transfer formulation
C_1	-	Constant in Woschni's heat transfer formulation
C_2	-	Constant in Woschni's heat transfer formulation
c_f	-	Long-range term of the general density variable $f(\bar{x},t)$ in the conservation equations

c_{HR}	-	Correction factor of the 3D-CFD-heat-release
c_{HR_lg}	-	Correction factor of the 3D-CFD-heat-release during the early flame development (ignition model)
c_{HT}	-	Correction factor of the 3D-CFD-heat-transfer
C_{Inj}	-	Discharge coefficient of the gas injector
Co	-	Courant number
c_p	J/kgK	Specific heat at constant pressure (thermal term)
C_T	-	Correction factor of the heat-transfer coefficient and the temperature difference (gas-wall) in the calculation of the 3D-CFD-heat-transfer
C_{tls}	-	Weighting factor for the turbulent length scale calculation in the combustion model
C_{turb}	-	Weighting factor for the turbulence calculation in the combustion model
$C_{\varepsilon 1-4}$	-	Empirical coefficients in the $\tilde{k} - \tilde{\varepsilon}$ standard turbulence model
C_μ	-	Coefficient in the turbulent viscosity formulation
D	μm	Size of the "fictive" gas droplets at injector orifice
D_i^M	m^2/s	Diffusion coefficient of species i due to a concentration gradient
$D_{i,T}^M$	m^2/s	Turbulent diffusion coefficient of species i
D_i^T	kg/mKs	Thermal diffusion coefficient of species i due to a temperature gradient
e	J/kg	Total specific energy
e	m^3/s^2	Turbulent energy density
f	-	General density variable of conservation equations
F	-	General extensive variable of conservation equations
F_{res}	-	Coefficient in the laminar flame speed correlation
f_μ	-	Coefficient in the turbulent viscosity formulation
g	-	Geometric interpolation factor in differencing schemes
$\bar{g}$	m/s^2	Gravity acceleration
G	J/kg	Potential gravitational energy per unit mass
$\overline{G}^0$	J/mol	Molar Gibb energy (free energy)
g_i	J/mol	Molar Gibb energy (free energy) of a species i
h	J/kg	Specific enthalpy (thermal part)
h	J/kg	Specific enthalpy of burned gas (thermal part)
H	J	Enthalpy
h_i	J/kg	Specific enthalpy of species i (thermal part)

h_f	J/kg	Specific heat of formation (chemical part)
h_f	J/kg	Specific heat of formation of burned gas (chemical part)
$h_{f,F}$	J/kg	Specific heat of formation of fuel (chemical part)
$h_{f,Fresh}$	J/kg	Specific heat of formation of fresh gas (chemical part)
$h_{f,i}$	J/kg	Specific heat of formation of species i (chemical part)
$h'_{f,i}$	J/kg	Specific molar heat of formation of species i (chemical part)
h_{HR}	J/kg	Specific oxidation heat-release
h_{LHV}	J/kg	Fuel lower heating-value
h_{tc}	J/kg	Specific thermo-chemical enthalpy
h_{tc}	J/kg	Specific thermo-chemical enthalpy of burned gas
$h_{tc,i}$	J/kg	Specific thermo-chemical enthalpy of species i
$\overline{\overline{I}}$	-	Unit matrix
$imep$	bar	Indicated mean effective pressure
I_0	-	Strain factor of the laminar burning speed
isfc	g/kWh	Indicated specific fuel consumption
$\vec{j}_i$	kg/m^{2}s	Diffusion mass flux of species i
$\vec{j}_i^{\,d}$	kg/m^{2}s	Ordinary diffusion mass flux of species i due to a concentration gradient
$\vec{j}_{i,T}^{\,d}$	kg/m^{2}s	Turbulent diffusion mass flux of species i
$\vec{j}_i^{\,P}$	kg/m^{2}s	Pressure diffusion mass flux of species i due to a pressure gradient
$\vec{j}_i^{\,T}$	kg/m^{2}s	Thermal diffusion mass flux of species i due to a temperature gradient
$\vec{j}_q$	W/m^2	Molecular heat flux
$\vec{j}_q^{\,c}$	W/m^2	Heat conduction
$\vec{j}_q^{\,d}$	W/m^2	Diffusion heat flux due to diffusion mass flux $\vec{j}_i$ of each species i
$\vec{j}_q^{\,D}$	W/m^2	Heat flux due to Dufour-effect
k	m^2/s^2	Turbulent kinetic energy (TKE)
$\overline{k}$	m^2/s^2	Relevant turbulent kinetic energy (TKE) in the phenomenological 3D-CFD-heat-transfer model
$k^{(b)}$	-	Reaction speed coefficient of the backward reaction
$k^{(f)}$	-	Reaction speed coefficient of the forward reaction
$\overline{k}_{0D}$	m^2/s^2	Average turbulent kinetic energy in the cylinder (zero- or quasi-dimensional approach)
K	-	Flame wrinkling factor $(K - S_T/S_L)$

K_{1_Step}	-	Equilibrium constant of the one-step oxidation (water-gas reaction)
K_c	-	Molar equilibrium constants
K_p	-	Equilibrium constants of the partial pressures
l	m	Characteristic length in the Re-Nu-correlation
$L_{3D-Mesh}$	m	Discretized length of the 3D-CFD-domain of the intake system
l_D	-	Averaged cell-discretization-length of the mesh
l	m	Turbulent length scale
l_g	m	Characteristic length scale of the cylinder
l_k	m	Kolmogorov turbulent length scale
l_l	m	Turbulent integral length scale
L_{min}	kg/kg	Stoichiometric minimum air requirement
l_s	m	In large-eddy simulation (LES), turbulent length scale switch from DNS solution to isotropic turbulence modeling
l_{Taylor}	m	Taylor's Turbulent length scale
m	kg	Mass
m	-	Coefficient in the Wiebe combustion function
m	-	Exponent in the Re-Nu correlation
m	-	H-Atom number in fuel composition - $C_nH_mO_rN_q$
$\overline{M}$	kg/kmol	Molar mass of the gas mixture
m_B	kg	Mass in the burned zone in the cylinder
M_{EGR}	kg/kmol	Molar mass of the residual gas
m_f	kg	Mass of the flame front in the cylinder
m_F	kg	Fuel mass trapped in the cylinder
M_F	kg/kmol	Molar mass of fuel
m_{F_U}	kg	Fuel mass in the unburned zone of the cylinder
m_i	kg/kg	Mass of species i
M_i	kg/kmol	Molar mass of species i
m_{Inj_F}	kg	Fuel injected mass
m_U	kg	Mass in the unburned zone in cylinder
M_U	kg/kmol	Molar mass of the unburned gas
n	rev/min	Engine speed
n	kmol	Total number of moles in the gas mixture
n	-	Exponent in Re-Nu correlation
n	-	Polytropic compression exponent
n	-	C-Atom number in the fuel composition - $C_nH_mO_rN_q$

$\vec{n}$	-	Normal vector unit
N_{Cells}	-	Total number of cells in the mesh of the combustion chamber
N_{B_Ig}	-	Total number of cells in the mesh involved in the early flame development (ignition model)
n_{EGR}	kmol	Number of moles of the residual gas
n_i	kmol	Number of moles of species i in the gas mixture
N_i	-	Total number of species in the gas mixture
n_U	kmol	Number of moles of the unburned gas
Nu	-	Nusselt Number
p	Pa, bar	Pressure
$\overline{\overline{P}}$	N/m^2	Stress tensor
p_{cr}	Pa	Critical pressure
p_i	Pa	Partial pressure of species i
Pr	-	Prandtl Number
q	-	N-Atom number in fuel composition - $C_nH_mO_rN_q$
$\dot{q}$	W/m^2	Wall heat-transfer flux
Q_B	J	Fuel heat-release energy
Q_{B_Ox}	J	Oxidation fuel heat-release
$Q_{Cooling}$	J	Engine cooling energy
Q_{Env}	J	Convective and radiation heat transfer between engine and environment
Q_{Ex}	J	Exhaust gas energy
Q_W	J	Cylinder wall heat-transfer
q_r	J/kgs	Specific energy long-range term due to radiation or magnetic fields
r	-	O-Atom number in fuel composition - $C_nH_mO_rN_q$
$\Re$	J/kmolK	Universal gas constant
R	J/kgK	Real gas constant
Re	-	Reynolds Number
$\widetilde{r}_B$	kg/m^{3}s	Reaction rate of burned gas
r_K	m	Flame kernel radius (SI-engines)
s_f	-	Production or sink term of the general density variable $f(\vec{x},t)$ in the conservation equations
s'_i	J/kmolK	Specific molar entropy of species i
S_L	m/s	Laminar flame speed (relative speed)
S_P	m/s	Instantaneous piston speed

$\bar{S}_P$	m/s	Mean piston speed
S_T	m/s	Turbulent flame speed (relative speed)
t	s	Time
T	K	Temperature
$\bar{T}$	K	Characteristic temperature in the Re-Nu-correlation
$T_{B\_ad,j}$	K	Adiabatic flame temperature in the 3D-CFD-cell j
T_B	K	Temperature of the burned zone in the cylinder
T_C	K	Temperature in a 3D-CFD-wall-cell
T_{cr}	K	Critical temperature
T_{freeze}	K	Temperature at "frozen state" chemical equilibrium
T_G	K	Gas temperature in the wall heat-transfer calculation
t_{IP}	s	Time at the ignition point
T_P	K	Temperature in the central node of a 3D-CFD-cell
T_U	K	Temperature of the unburned zone in the cylinder
T_W	K	Wall temperature
u	J/kg	Specific internal energy
u	m/s	Instantaneous fluid velocity component (scalar)
u'	m/s	Turbulent velocity
U	J	Internal energy
$\bar{U}$	m/s	Ensemble-averaged mean velocity component
$\hat{U}$	m/s	Cycle-by-cycle variation in mean velocity
u_f	m/s	Absolute flame front speed
u_T	m/s	Turbulent term of the relative flame front speed
u_τ	m/s	Shear friction velocity component (near-wall region)
$\bar{v}$	m/s	Fluid velocity (vectorial)
V	m³	Volume
V	m³	Cylinder volume
V_B	m³	Volume of the burned zone in the cylinder
V_{B_lg}	m³	Volume of the burned zone in the cylinder during the early flame development (ignition model)
V_C	m³	Cylinder clearance volume
$V_{\bar{C}}$	m³	Volume of a discrete element (cell)
V_D	m³	Cylinder displaced volume
$\bar{v}_i$	m/s	Velocity of species i
$\bar{V}_i$	m/s	Diffusion velocity of species i

V_{Inj}	m/s	Gas injection velocity
v_T	m/s	Fluid velocity term in the flame propagation model (combustion model)
V_U	m³	Volume of the unburned zone in the cylinder
w	m/s	Characteristic velocity in the Re-Nu-correlation
$w_{Air\_B,j}$	kg/kg	Mass fraction of air in the 3D-CFD-cell j that has previously produced the burned gas
$w_{Air\_U,j}$	kg/kg	Mass fraction of fresh air in the 3D-CFD-cell j
W	-	Coefficient in the laminar flame speed correlation
w_B	kg/kg	Mass fraction of burned gas in the cylinder
$w_{B,j}$	kg/kg	Mass fraction of burned gas in the 3D-CFD-cell j
W_{Eff}	J	Effective engine work
$w_{EGR,j}$	kg/kg	Mass fraction of EGR in the 3D-CFD-cell j
$w_{EGR\_Air\_U,j}$	kg/kg	Mass fraction of air in the 3D-CFD-cell j that has previously produced EGR (burned gas of the previous operating cycle)
$w_{EGR\_F\_U,j}$	kg/kg	Mass fraction of vaporized fuel in the 3D-CFD-cell j that has previously produced EGR (burned gas of the previous operating cycle)
w_{EGR_U}	kg/kg	Mass fraction of residual gas in the unburned zone
w_F	kg/kg	Mass fraction of fuel trapped in the cylinder
$w_{F\_U,j}$	kg/kg	Mass fraction of fresh vaporized fuel in the 3D-CFD-cell j
$w_{F\_B,j}$	kg/kg	Mass fraction of vaporized fuel in the 3D-CFD-cell j that has previously produced burned gas
$w_{Fresh,j}$	kg/kg	Mass fraction of fresh charge in the 3D-CFD-cell j
w_i	kg/kg	Mass fraction of species i
W_I	J	Indicated piston work
w_{F_U}	kg/kg	Mass fraction of fuel in the unburned zone
W_R	J	Engine friction losses
$\bar{x}$	m	Position vector
x_{EGR_U}	kmol/kmol	Mole fraction of residual gas in the unburned zone
x_i	kmol/kmol	Mole fraction of species i
y	mm	Distance to the wall
y^+	-	Dimensionless wall distance
Z	-	Coefficient in the laminar flame speed correlation

Greek Symbols

α	W/m^{2}K	Convective heat-transfer coefficient (gas side)
α	-	Coefficient in the laminar flame speed correlation
α	deg	Cone injection angle
α_1	W/m^{2}K	Convective phenomenological 3D-CFD-heat-transfer coefficient (gas side) during the working period
α_2	W/m^{2}K	Convective phenomenological 3D-CFD-heat-transfer coefficient (gas side) during the exchange period
β	-	Coefficient in the laminar flame speed correlation
χ_{Taylor}	-	Taylor's coefficient in turbulence modeling
Δ	-	Combustion term of Bargende's heat transfer formulation
$\Delta_R \overline{G}^0$	J/kmol	Change of Gibbs energy of a chemical reaction
Δt	s	Delta time-step of the 3D-CFD-simulation
$\Delta \varphi$	deg	Delta crank-angle of the 3D-CFD-simulation
ε	deg	Internal angle of the hollow injection cone
$\overline{\varepsilon}_{0D}$	m^2/s^3	Average turbulent kinetic energy dissipation velocity in the cylinder (zero- or quasi-dimensional approach)
ϕ	-	Fuel/air equivalence ratio $\left(\phi = 1/\lambda \right)$
$\overline{\Phi}_f$	-	Flux of the general density variable $f(\vec{x}, t)$ in the conservation equations
Φ_m	-	Coefficient in the laminar flame speed correlation
φ	deg	Crank angle
φ_C	deg	Crank angle at the end of combustion
η	-	Coefficient in the laminar flame speed correlation
η_C	-	Combustion efficiency (imperfect combustion)
η_{HR}	-	Combustion conversion efficiency (incomplete combustion)
η_V	-	Volumetric efficiency
κ	-	Specific heat ration $\kappa = c_p/c_v$
λ	-	Relative air/fuel ratio $\left(\lambda = 1/\phi \right),$
λ	W/mK	Thermal conductivity
$\lambda_{B,j}$	-	Lambda of burned gas in the 3D-CFD-cell j
$\lambda_{EGR,j}$	-	Lambda of EGR in the 3D-CFD-cell j
$\lambda_{Fresh,j}$	-	Lambda of fresh gas in the 3D-CFD-cell j
λ_T	W/mK	Turbulent thermal conductivity
μ	kg/ms	Dynamic viscosity

μ_T	kg/ms	Turbulent dynamic viscosity
$\overline{\overline{\Pi}}$	N/m^2	Shear-stress tensor
$\overline{\overline{\Pi}}_T$	N/m^2	Turbulent shear-stress tensor
ρ	kg/m^3	Density
ρ_U	kg/m^3	Averaged density in the unburned zone of the cylinder
σ_k	-	Empirical coefficient in the $\tilde{k} - \tilde{\varepsilon}$ standard turbulence model
σ_ε	-	Empirical coefficient in the $\tilde{k} - \tilde{\varepsilon}$ standard turbulence model
ς	-	Coefficient in the laminar flame speed correlation
τ_l	s	Combustion burn-up time in the quasi-dimensional approach
ω_i	$kmol/m^3s$	Mole formation rate of species i due to chemical reactions

Subscripts and Abbreviations

0	Standard thermodynamic conditions
0	Simulation start (initial conditions)
1D	One dimensional
3D	Three dimensional
ad	Adiabatic
B	Burned
BB	Blow-by
BD	Blended differencing scheme
BDC	Bottom dead center
CA	Crank angle
CFD	Computational fluid dynamics
CI	Compression ignition
CD	Central differencing scheme
CNG	Compressed natural gas
Corr.	Correction
CPU	Central processing unit
corr.	Correction
cyl.	Cylinder
DI	Direct injection
DNS	Direct numerical simulation
E	Exhaust

ECU	Electronic control unit
Eff	Effective
EGR	Exhaust gas recirculation (residual gas)
Env	Environment
EVC	Exhaust valve close
EVO	Exhaust valve open
Exp	Experiments
F	Fuel
FBDC	Fire bottom dead center
FEM	Finite elements method
FF	Flame front
FTDC	Fire top dead center
HCCI	Homogenous charge compression-ignition
HOF	Heat of formation
HR	Heat release
HT	Heat transfer
I	Inlet
IC	Internal combustion
Ig	Ignition
Ind	Indicated
IP	Ignition point
IVC	Intake valve close
IVO	Intake valve open
LES	Large eddy simulation
LHV	Lower heating value
LPG	Liquefied petroleum gas
LRN	Low Reynolds number
LTC	Low temperature combustion
LUD	Linear upwind differencing scheme
MARS	Monotone advection and reconstruction scheme (differencing practice)
MBS	Multi-body system
Ox.	Oxidation
PC	Personal computer
PH	Phenomenological
PISO	Pressure implicit split operator

QUICK	Quadratic upstream interpolation of convective kinematics differencing scheme
Ref	Reference
RON	Research octane number
SFCD	Self-filtered central differencing scheme
SI	Spark ignition
SOI	Start of injection
SP	Spark plug
TDC	Top dead center
TKE	Turbulent kinetic energy
tls	Turbulent length scale
U	Unburned
UD	Upwind differencing scheme
UHC	Unburned hydrocarbons
VVT	Variable valve timing
W	Wall
WF	3D-CFD Wall function (heat-transfer calculation)
WOT	Wide open throttle
WP	Real working-process analysis

1

Introduction

1.1 Society and Transportation

Society is a dynamic being and transportation is a key factor for fulfilling the human needs: going to work, shopping, bringing the children to school, travelling, delivering goods, etc., are just a short list of activities related to mobility, which plays an important part in economic growth, globalization and, on the other side, unfortunately also causes a certain environmental impact. Among different means of transport, vehicles equipped with internal combustion engines represent the most relevant share in the global transportation and namely cars and motorcycles are a self-evident object in our daily activities.

1.2 The Fascination of Internal Combustion Engines

Since the end of the 19th century no other kind of propulsion has contributed to the personal transportation more than internal combustion engines. Thereby, without doubts, torque, power, sound and response to the gas pedal attribute a "personality" to the engine in a way that fascinates. Internal combustion engines are machines that are for sure far away to be the "perfect propulsion device" but in the human imagination and subconscious no other one can offer more feeling. This is one more proof that the human being does not act rationally (symbiosis between man and car). A Ferrari's engine without its typical sound during acceleration (sometimes described by the Italian press as a *fortissimo of a full orchestra*) would not attract anymore and a modern 3-liters turbocharged diesel engine without its brutal torque output by driving on an alpine pass would be incredibly boring. There is no doubt, that these engine behaviors are not relevant in fulfilling the need of transportation and maybe somebody would like to prescribe the maximal number of allowed horse powers in a car and also how the car has to look or, even more drastically, that private traffic using cars equipped with internal combustion engines should be definitely prohibited and substituted by other means of transport. Apart from these isolated thoughts, cars and their internal combustion engines will still have a central role in the human wishes of mobility, freedom, adventure and technical challenge. Therefore car manufactures will always have to follow carefully the trends of the market by introducing new models that, first of all, have to fulfill the harder and harder legislation rules, then must meet rational and irrational customer expectations and finally have to find enough buyers ready to pay for the proposed technical solutions. In the past, several times, it has been experienced by many manufactures that cars with too many limitations and without a certain "sex-appeal" are destined to become a fiasco; therefore we must be conscious that fascination, expectations and affordability will continue driving our life.

1.3   Internal Combustion Engines and Sustainable Transportation

As introduced before, everybody in the future will continue having the right to buy the car that better meets his expectations and wishes and for sure this decision process still will be, at least in part, an irrational process. What has changed in the last years and will continuously become harder in the future are the limitations on the environmental impact of transportation in general and in particular of private transportation. This welcomed, positive and irreplaceable process towards a real sustainable mobility will dictate the future development steps in the automotive sector.

1.3.1   Development Targets of Internal Combustion Engines in the Past

At the beginning of the history of internal combustion engines the development priorities were principally the increasing of power and engine speed. In that ages cars were very exclusive products; the maximal car speed was the driving idea during the development process, therefore cars with an engine displacement over 10 liters (up to a maximum of 21.5 liters for the 1912 Benz 82/200) were not rarely produced. The Ford Model T determined 1908 as the historic year in which the automobile came into popular usage [1]. It is generally regarded as the first affordable automobile, the car that "put America on wheels". The engine became smaller (still 2.9 liters), simplified for mass production, and the ratio power to engine weight gained in importance.

After WWII, during the years, development priorities became principally the technical reliability, then engine smoothness, displacement reduction towards costs minimization (at least in Western Europe) and noise reduction (usually achieved only by better mufflers).

Not until the fifties the effect of air pollution on health became so evident that it was not possible to be ignored anymore [2]. As early as 1953, Los Angeles County Supervisor Kenneth Hahn inquired of Detroit car manufactures whether research was being conducted to eliminate emissions. The response was vague. With the threat of mandatory federal regulations, the automotive industry began to install crankcase blow-by devices on their cars. This was a significant advance because crankcase blow-by produced 25 percent of the engine hydrocarbon-emissions. This equipment became mandatory on all cars sold in California beginning with the 1963 models. After this first step, in 1966 California prescribed hydrocarbon exhaust-control devices up to 1975 when catalytic converters were imposed on all new cars.

In the seventies, after two world oil crisis that caused a dramatic increase of the oil price, the reduction of the fuel consumption gained drastically in importance. Especially in Western Europe and in Japan many tests were carried out towards increasing of engine efficiency: from lean combustion, improved combustion chamber shapes, etc., up to rudimental start-stop strategies at the beginning of the eighties. In particular it was the appearance of electronics that permitted to develop the first ECUs so that, first of all, the dosage of the fuel mixture became very accurate before more and more complex engine map applications allowed a more efficient engine operation.

In the time range between the eighties and the end of the 20th century, internal combustion engines have been intensively further improved. The main development targets for gasoline engines have been technical reliability, exhaust emission reduction, noise reduction and very high engine smoothness. Because of a remarkable increasing of the average weight of the cars caused by safety regulations and market trends and in order to continuously offer more performing vehicles the displacement and number of cylinders of the installed engines has also been increased. So a great part of the improved energy efficiency of gasoline engines has been sacrificed, i.e. only a moderate reduction of fuel consumption has been achieved in these years. In the same time the development process of diesel engines has been decisively more dynamic. Old, smoking, heavy, natural aspirated pre-chamber diesel-engine with very low specific power have been substituted by powerful, smaller, cleaner, turbocharged direct-injected diesel engines with a very high efficiency. In Western Europe, this new generation of diesel engines has drastically increased the market share within short time: more than 40% in average and up to 75% in the market segment of SUV and luxury class. Especially in these classes where the weight of the car is higher and engine displacement is at about 3 liters the advantages in term of fuel saving and CO_2-emissions are unbeatably by the diesel engine in comparison to an equivalent gasoline one.

1.3.2   The Role of Alternative Engine Concepts

Since the beginnings of the automotive mobility, the interest in finding competitive alternative concepts to internal combustion engines has always been very high. Many different solutions have been tested, but actually only electric engines represent the only possible alternative propulsion for commercialization. Electric cars are commonly powered by on-board battery packs but also sophisticated devices (fuel cells) for conversion of chemical energy of the fuels (at best hydrogen) into electricity can be used for energy storage. During the last years many prototype vehicles equipped with fuel cells have been tested, but at the moment, principally prohibitive production costs and an inadequate hydrogen refueling infrastructure have drastically reduced the expectations of commercialization in the near future.

Today, first of all, the density of energy stored in conventional batteries is too low for ensuring an acceptable autonomy of the vehicle, furthermore the cost of the battery pack and its life-time it is not competitive to internal combustion engines. Therefore the future of battery-equiped electric vehicles depends primarily on the cost and availability of batteries with high energy densities, power density and long life, as all other aspects such as motors, motor controllers and chargers are fairly mature and cost-competitive with internal combustion engine components.

Innovative but for automotive mobility less experienced Li-ion, Li-poly and zinc-air batteries have recently demonstrated potentiality of energy densities high enough to deliver acceptable autonomy and recharge time. Therefore the market share of electric vehicles in the future depends on the development of battery technology in term of costs, reliability, energy density and infrastructure for recharging. Hence, today precise prognosis for the far future are very difficult, but for the next future (up to 2020) a market share to some extent can be expected only in the sector of city-cars.

1.3.3   Development Targets of Internal Combustion Engines in the Future

In the last years the internal combustion engine has come extremely under fire. More and more strict exhaust emission regulations (especially for diesel engines), the necessity of a remarkable reduction of CO_2-emissions and a certain concurrence from alternative engine concepts (still in the prototype step) have begun the discussion about the future of internal combustion engines.

But as introduced before, there are no scenarios that see a remarkable electrification of the transport sector in the near future (up to 2020). Also optimistic studies banish the electrification to niches (city cars), i.e. internal combustion engines will still play the dominant role. Nevertheless in order to face the challenge on a long term, new development targets are required, i.e. only with "courageous" development strategies the full potentiality of the internal combustion engines can be better exploited.

The traditional internal combustion engine, i.e. a natural aspirated gasoline engine with large displacement and port injection is definitely an obsolescent concept that will be pursued no longer. In the last years a drastic change of design strategies has already started. Under these conditions engine manufacturers are ready to make decisions and introduce solutions that just few years ago would have "shocked" the market. The main topics of future engine development are briefly discussed here.

1.3.3.1 General Improvement of actual Solutions

Surely, future development strategies will require the introduction of innovative solutions; nevertheless a general improvement of actual solutions in combination with the innovative ones will increase the contribution to efficiency enhancement. Among others, friction reduction, adaptive valve timing devices up to a full VVT-engine and improvements of fuel injection systems, combustion process, engine mapping, thermo-management, start-stop strategies, etc. represent just few of many topics that can influence the engine efficiency.

1.3.3.2 Downsizing and Turbo-Charging

The engine of the future has a reduced displacement and number of cylinders (commonly from 2 up to 4 cylinders, more cylinders, if at all, only in the luxury class, sport cars and large SUVs). Sophisticated turbo-charging systems are required for replacing the power losses due to displacement reduction. The result is an engine with the same power of a traditional natural aspirated engine but with a better torque curve, less fuel consumption (up to 20% and even more in urban tests), reduced size and, for this reason, less engine weight.

1.3.3.3 Hybridization

Basically, a hybrid electric vehicle combines an internal combustion engine and an electric motor powered by batteries, merging the best features of both propulsion devices. This expensive combination allows the electric motor to help the conventional engine operating more efficiently by providing additional power to the car transmission. In addition a hybrid car allows recovering a part of the kinetic energy of the vehicle during braking instead of dissipating it. Especially in case of start-stop conditions, low vehicle speed, accelerations from low engine speed, etc., the "generous" torque output of the electric motor supports the "modest" one of the internal combustion engine (even more evident by downsized turbo-charged engines during the turbo-lag phase). In these situations the fuel saving is at the maximum. Under other conditions, e.g. constant vehicle speed on a highway, the hybridization is useless and the car is still negatively affected by a remarkable increase of weight.

1.3.3.4   Development of Innovative Combustion Solutions

In the last years several innovative combustion solutions for gasoline and diesel engines towards a promising reduction of fuel consumption and exhaust emissions have been investigated. The most interesting solution here is represented by the HCCI-approach (Homogeneous charge compression ignition) [3]. In this approach, under certain engine operating conditions (middle speed range and from idle up to partial load) a well-homogenous mixture of fuel and air is compressed to the point of auto-ignition, i.e. an HCCI-engine has characteristics of the two most popular forms of combustion used in internal combustion engines: homogeneous charge spark ignition (gasoline engines) and stratified charge compression ignition (diesel engines). Since HCCI-engines can operate with leaner mixtures than spark ignition engines, the peak temperatures during combustion are lower. The low peak temperatures prevent the formation of NO_x. This leads to NO_x emissions at levels far less than those found in traditional engines.

The difficulties in controlling both the HCCI-combustion and the switching procedure from HCCI-operation to a traditional combustion process (e.g. at full load) are the major hurdles that have to be eliminated in the future until commercialization. In a typical gasoline engine, a spark is used to ignite the pre-mixed fuel and air. In diesel engines, combustion begins when the fuel is injected into compressed air. In both cases, the timing of combustion is explicitly controlled. In an HCCI-engine, however, the homogeneous mixture of fuel and air is compressed and combustion begins whenever the appropriate conditions are reached. This means that there is no well-defined combustion initiation that can be directly controlled. This still leads to uncontrolled in-cylinder peak pressures that may cause a higher engine wear up to irreparable damages.

1.3.3.5   Alternative Fuels

In order to reduce CO_2-emissions and decrease the dependency on import of crude oil, in the last years, the interest in alternative fuel has remarkably gained in importance. Alternative fuels can be divided into three main categories: *fossil fuels* (Methane and LPG), *bio-fuels* (bio-diesel, bio-alcohol, bio-gas, etc.) and *hydrogen*.

Alternative fossil Fuels and Hydrogen

Alternative fossil fuels, in particular methane (CH_4), thanks to the higher H/C ratio in the fuel molecules allow a relevant CO_2-reduction during combustion (about 20% for methane up to 100% for hydrogen, which can be considered as the ideal "extrapolation" of the fossil fuels). Vehicles equipped with internal combustion engines running with LPG or methane fuel are a well-tried alternative to conventional ones and continuously gain market shares. Hydrogen vehicles because of high development and production costs and an inadequate refueling infrastructure, at the moment, have not left the prototype phase; hence a commercialization for mass production in the next future is not expected.

Bio-Fuels

Bio-fuels [4], mainly derived from plant materials, theoretically represent a CO_2-neutral alternative to common fossil fuels, which absolutely does not mean combustion without CO_2 emission. In comparison to common fossil fuels derived from biological organisms which lived millions of years ago, bio-fuels are derived from organisms which grow and are harvested today, before their biological material is converted into fuel. In both cases the CO_2 output during combustion is the same, but in case of bio-fuels the contribute of CO_2 reduction during the life of the involved biological organisms can be counted into the balance, so that at the end a CO_2-neutrality can be assumed. This is the theory, the reality is not so clear. The issue of discussion is the indirect impact of each bio-fuel source. Cutting down a rainforest releases a massive quantity of carbon which otherwise would have remained absorbed in the trees. Further the loss of rainforest means much less global forest to convert carbon dioxide into oxygen.

A recent study by the United Nations Energy Program comes namely to the conclusion, that bio-fuels should be considered climate-friendly (or not) based on the source. Whether the bio-fuel was made from a crop grown specifically to create that fuel after deforestation, or whether it came from crop residues, this has very different implications. In the first case the CO_2 balance can look worse than in case of consumption of common fossil fuels. The report also introduces the term of "acreage requirements" for different energy sources. For example the land required to grow bio-fuels can be enormous, while much less land is required to generate an equivalent amount of energy from wind or solar.

Bio-diesel is the most common bio-fuel in Europe. It is produced from oils or fats using transesterification and is a liquid similar in composition to fossil/mineral diesel. Bio-diesel can be used as a fuel for vehicles in its pure form, but it is commonly used as a diesel additive (in 2009 on the German market with a blending percentage of 5,25%).

Bio-ethanol is an alcohol made by fermenting the sugar components of plant materials and it is made mostly from sugar and starch crops. With advanced technology being developed, cellulosic biomass, such as trees and grasses, are also used as feed-stocks for ethanol production. Bio-ethanol can be used as a fuel for vehicles in its pure form, but it is usually used as a gasoline additive for increasing the octane number and improving vehicle exhaust emissions. Bio-ethanol is widely used in the USA, Brazil and North Europe.

Biogas is produced by the process of anaerobic digestion of organic material by anaerobes. It can be produced either from biodegradable waste materials or by the use of energy crops fed into anaerobic digesters to supplement gas yields. The actual consumption of biogas in the automotive sector is remarkably under the level of bio-diesel and bio-ethanol, but for the future a relevant demand increasing has to be expected.

1.4   How to Face the Complexity of Future Internal-Combustion-Engines

The engine of the future is lightweight, small, turbocharged, probably with an unconventional combustion strategy, silent, with extremely low exhaust emissions (also at cold start), very low fuel consumption and CO_2 emission, high specific power, "vigorous" torque at low engine speed, long-life reliability and, maybe the most difficult target, it must be still affordable.

Now the question is how to organize the future engine development strategy in order to face the increasing complexity and still keep the development and production costs in an affordable business plan. In engine development the process that permits to turn promising ideas into a final concept is very complicated and resource-expensive. This process has to take the following steps into account:

- technical and economical evaluation of the potentiality of an idea

- prototype realization

- prototype verification and improvement

- final engine realization

- final engine verification and improvement

- final engine control.

For these tasks several different development tools are at disposal. These tools can be divided, first of all, into two main categories: experimental (test bench and labor investigation) and theoretical (calculation and simulation analysis) approach, respectively. Each tool of these main categories differs in terms of application range, level of details, predictability, resource consumption, etc., so that at the end the spectrum of available development tools is very wide. Evidently the engine design process does not rely only on one development tool and especially in the future any task will require a closer integration of them.

It is one of the principal aims of this work to deal with this issue. In the next chapters a deeper analysis of the development tasks will be reported. In particular an analysis of the role of simulating in the engine design process and the introduction of an innovative and promising simulation solution will be discussed.

2

Simulation of Internal Combustion Engines

In engineering, simulating has come to mean the development and usage of appropriate formulations that permit critical processes, which take place inside the object of interest, to be analyzed.

Since the early developments of the internal combustion engine in the 19th century, e.g., the simulation of the operating cycle has remarkably contributed to increase the engine performance by estimating potentialities, limitations and practicability of different concepts [5-9].

Nicolaus A. Otto (1832-1891) as well, used the simulation to calculate the expected indicated work of different operating-cycle concepts before prototyping them [7]. These simulations based on gas-law equations permitted to identify a great advantage of combustion in a compressed fuel-air mixture, instead of atmospheric pressure. This was the key idea for an essential evolution of internal combustion engines.

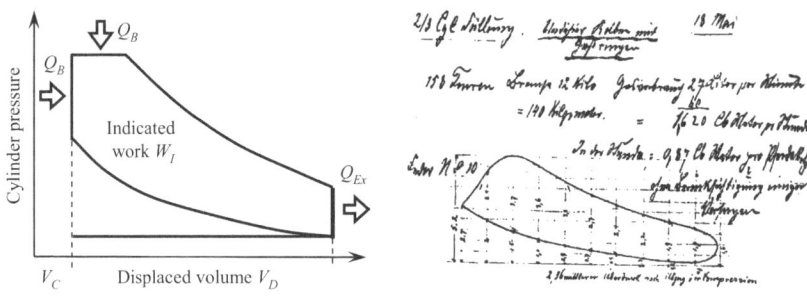

Figure 2.1: *Ideal p-V diagram of the engine operating-cycle proposed by Nicolaus Otto.*

Figure 2.2: *Nicolaus Otto's first experimental indicating diagram (18th May 1876).*

In order to realize this solution Nicolaus A. Otto, in 1867, was the first to propose and then patent an engine cycle with a four-stroke cycle: intake, compression, expansion and exhaust (Otto's cycle – see Figure 2.1). Experimental pressure data of the gas in the cylinder over the operating cycle of Otto's engine prototype confirmed the expectation from the simulation results (see Figure 2.2). In comparison to the first generation of marketable internal combustion engines with combustion at atmospheric pressure (efficiency at best about 5 percent) Otto's engine permitted to achieve both a thermal efficiency up to 11 percent and an enormous reduction in engine weight and volume [5]. This was the ancestor of contemporary automotive engines.

2.1   Simulation towards Virtual Engine Development

During the years engine simulation has continuously gained in reliance but it was mainly in the last two decades that the rapid increase of computer performance has faced the rising complexity of the engine design process by providing solutions supported by sophisticated simulation programs. At the present time the application of simulation programs that provide a reliable "virtual engine development" represents one of the greatest challenges in the development of future internal combustion engines [10,11].

The role of the simulation in the development process of internal combustion engines and the definition of virtual engine development are still not well defined. Thus engine manufacturers have quite different expectations regarding the considered aspect. For developers the emphasis is most certainly towards the gain of knowledge. Designers aim to exploit the possibilities by means of calculations, but also the limits of their design ideas even during the phase of concept finding. Test engineers desire to find explanations for measured phenomena and to be inspired to seek improvements. The management, on the other hand, appreciates rather the reduction of the technical risk involved with a new development, a possible reduction of development time and investment ("engine test stands are more costly than workstations") and last but not least reductions in both the development budget, as well as subsequently the manufacturing costs during production.

2.1.1   One Tool for the Simulation of the Entire Engine?

Internal combustion engines are extremely complex machines where detailed and accurate analyses of both the thermodynamic processes occurring in the fluids (charge, cooling, lubrication, etc.) and the stresses, strains and thermal loads of each mechanical part are particularly ambitious. Most of the processes are unsteady, highly spatial occurring, reactive and in their behaviors in part still undiscovered, therefore at the "state-of-the-art" of today's simulations tools there is no tool able to comprehensively reproduce the functionality of an internal combustion engine in its integrity.

During the years numerous simulations tools with different degrees of complexity, versatility, predictive capability and time response have been developed in order to analyze targeted topics which have been recognized as relevant for the operating and then consequently for the improvement of internal combustion engines. Each of these simulation tools finds its application in the engine design process in combination with other ones depending, among other things, on the priorities set in the development process, on the experience of the developers with certain tools (each developer, in part subjectively, sees the limitation and the applicability of each tool in a different way) and last but not least on the resources available. These are the main reasons why the demand and the combination of simulation tools in relation to experimental investigations vary strongly not only among different manufacturers but also within each department involved in the engine design process.

Here a brief description of the most important available simulation tools for engine design process aims to introduce their peculiarities in terms of application, performance and result reliability (level of maturity). In a first step, these tools can be generally divided into two main categories: for mechanical and engine operating cycle analysis, respectively.

2.1.1.1   Mechanical Numerical Analysis

Mechanical numerical analyses are performed by two main categories of simulation tools: Multi body system (MBS) and finite elements methods (FEM) simulations [10].

Multi-Body System (MBS)

In the multi-body system (MBS) approach mechanical components are modeled as a sum of relatively simple spring-mass systems that are linked among each other within a dynamic structure. Thanks to this coarse discretization, a MBS simulation allows in a first step to calculate the kinematic and dynamic of the object of interest with moderate computing times. In a second step, starting from an estimation of the structural properties of the components involved (quite difficult for complex geometries) it is possible to calculate the strains and deformations in each part of the spring-mass system. For example, multi-body simulations are reliable in the determination of both the bending forces and the maximal torsion in a crank shaft over the whole range of engine speed and load, so that a preliminary design of the crank shaft can be performed within a short development time.

MBS is a very established approach for mechanical analysis of the entire engine (often the entire transmission is also included in the analysis) that permits first of all a general estimation of the mechanical stresses in each part, then to investigate eventual weak points of the engine design process and finally to carry out modifications and evolution. During the years additional tools have been implemented in MBS simulations which allow to estimate vibrations, acoustic, etc. Depending on the degree of the discretization, assumptions, calibration and complexity of the problem these additional tools have reached different levels of maturity, nevertheless their implementation is always worth for preliminary investigations.

Finite Elements Method (FEM)

In the finite elements method (FEM) approach, in a similar way like in molecular structures but on a higher spatial scale, the mechanical component of interest is finely discretized into a grid of elemental entities. Here the grid reproduces in details the weight and form of the original part with the same mechanical and structural properties (inertia moments, stiffness, etc.). This time expensive calculation allows a punctual calculation of the local stresses and deformations in each part of the mechanical component also in case of very complex geometries.

Thanks to this approach a design refinement of each engine component can be performed. For example, the development of a crank shaft for a race engine for which, starting from a base design, weight reduction in combination with an increased stiffness are required, FEM calculations are an irreplaceable approach for a successful task. A FEM simulation of the entire engine is not recommendable because not only the high computation time but also limitations and the imprecision of few contact algorithms (e.g. hydraulic contact between piston and cylinder liner), the determination of the thermal load of the engine parts, etc., would reduce the benefit of such a simulation. Therefore the application of FEM simulations is limited to the improvement of selected parts of the engines, where, without doubt, the reliability of the results is very high when the definition of the boundary conditions (mechanical and thermal) is accurate.

2.1.1.2   Engine Operating Cycle Analyses

Similar to mechanical analyses numerical investigations of the engine operating cycle can be performed by two main categories of simulation tools: real working-process analyses (WP) and three-dimensional fluid-dynamics simulations (3D-CFD). In engine design processes there are also simulations (1D-CFD) that do not below to a new category but they are just a combination of the two approaches introduced here. The following description of the simulation tools is only a very short introduction of this work because more detailed information will be reported in the next chapters.

Real Working-Process Analysis (WP)

The approach of the real working-process analysis is in nature thermodynamic and is limited to the evaluation of the engine operating cycle with a combustion chamber represented as an open thermodynamic system. This considers the changing of the thermodynamic state of the working fluid within the combustion chamber using implemented equations based on the energy conservation. Because it is not the aim to solve the fluid motion this kind of simulations are often called "zero-dimensional". The solution of the operating cycle requires additional information from the test bench (pressure profile) or assumptions over the combustion profile, i.e. because of this lack in predictability, this approach is actually rather an analysis or calculation tool than a simulation tool. The computational time is very short and the results reliability is after more than 40 years of development very high, even for innovative exchange process and combustion strategies. Hence real working-process analyses are an irreplaceable tool for the evaluation of the engine operating cycle as postprocessor of the pressure indicating system or recently even on-line at the test bench.

Three-Dimensional Fluid-Dynamics (3D-CFD)

The three-dimensional fluid-dynamics simulation is like FEM a no engine specific tool that performs the full analysis of the fluid motion independently from the complexity of the geometry. In this "multi-dimensional" approach the object of interest is finely discretized into a grid of finite volumes which reproduces the original shape. The solving of the conservation equations (mass, species mass fraction, momentum and energy) in any finite volume allows a detailed and comprehensive analysis of the reactive flow field within the engine. As the computing time is very high, the domain of the 3D-CFD-simulation is limited only to the parts of the engine that have either a complex geometry or where complex phenomena take place (usually the combustion chamber or the airbox where numerical investigations on fuel injection, combustion, etc. are required).

The limitation of the simulation domain requires the application of accurate boundary conditions, otherwise they will irremediably compromise the reliability and the predictability of the results of the 3D-CFD-simulation. This is one of the main critical points that will be widely discussed in Chapter 11. 3D-CFD-simulations are a quite recent approach in the engine development and due to the complexity in modeling the processes that take place in an engine, despite the potentiality, they have not reached an overall high level of maturity.

One-Dimensional Fluid-Dynamics (1D-CFD)

One-dimensional fluid-dynamics simulations are a combination of both the thermodynamic approach of the real working-process analysis for the cylinder and a simplified fluid dynamic simulation (one-dimensionally resolved along the flow direction in the pipes) for both the intake and exhaust system (other more complex components like the turbo charger are implemented as a "black box"). This simplification supported by an accurate model calibration with experimental data permits to analyze and predict the engine exchange process with good accuracy by remarkably reducing the computing time in comparison to 3D-CFD-simulations.

The real working-process analysis in combination with a predictive calculation of the exchange process gains in predictability, de facto, due to the absence of a validated model for the determination of the fluid motion within the combustion chamber, only the limitation in the setting of an adequate predictive combustion profile still remains. Recently, improvements in the development of so called quasi-dimensional combustion models [12-18] which take into account, among other things, few geometrical parameters of both the intake channel and the combustion chamber have permitted to realize a "half" predictive approach that recognizes a trend in the combustion profile. In the last ten years 1D-CFD-simulations have become a reliable tool in engine development and they are able to find improved solutions especially in turbo-charging layout, valve timings setting, general design of intake and exhaust track, etc., as far as the geometrical complexity of the investigated engine components can be adequately reduced or discretized into a layout of one-dimensional subparts.

2.1.2 The Future Challenge: an improved Integration of Simulation Tools

As introduced before an all-comprehensive tool for a reliable, detailed and predictive simulation of an internal combustion engine does not exist and will probably never exist. But there is another more realistic way represented by a selective coupling of different simulation tools depending on the different targeted tasks of the engine development. Trying to extend such coupling and linking of various simulation tools towards a comprehensive virtual engine development, a "hypothetical" scheme like that shown in Figure 2.3 can be imagined. Of course the pattern represents only a selection, further calculation tools could be added without problem. However, the listed tools already contain a very great number of man years in development by engineers, natural scientists and computer scientists. Under these aspects it is completely impossible to develop a kind of "great all-encompassing simulation tool". Such a project would not be successful, as neither time nor cost and last but not least qualified staff would be available in sufficient quantity. Thus, the "virtual engine" in the sense of a complete "virtual" development without test bench measurements does not exist and will – in all probability – never exist. But on the other hand the value of calculation and simulation in the development process of combustion engines is increasing continuously.

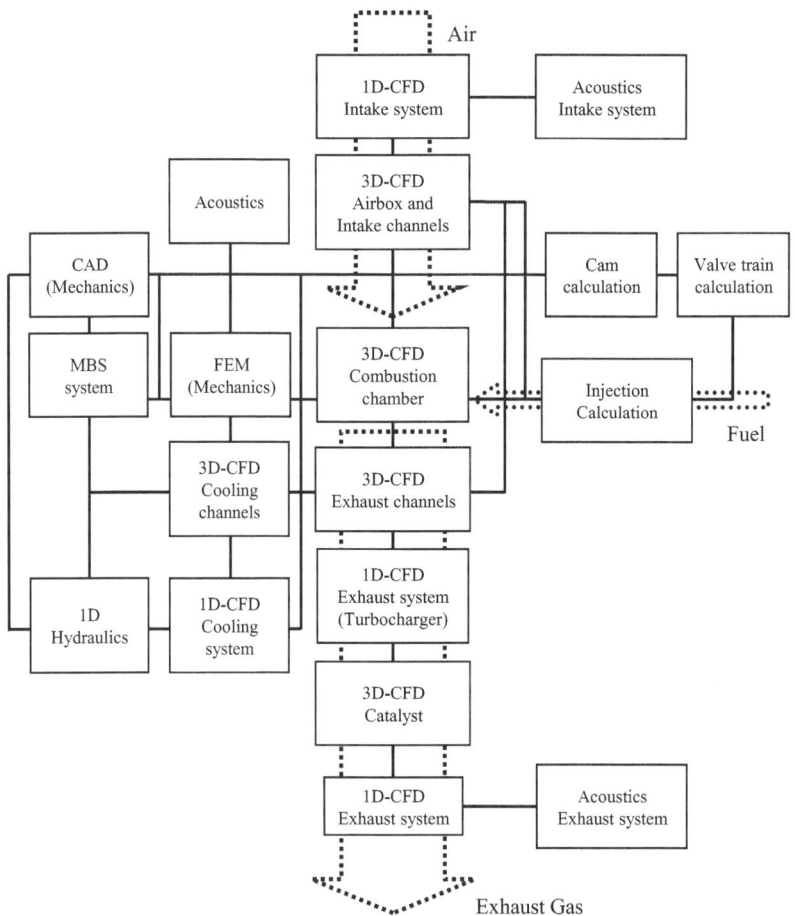

Figure 2.3: *Hypothetical integration map towards fully virtual engine development.*

The future definitely lies within the continued development of existing tools, the development of complementing tools and an intelligent coupling of them, so that the computational support in engine development can become more relevant. More complex technical solutions in the future that will permit great development potentials of combustion engines (see Chapter 1) would possibly not even be taken into consideration as series solutions without computational support.

Design, calculation, simulation and testing as the "main columns" of engine development will collaborate even more closely in the future [19], whereby this process must be handled with great care. Demands like "there must be less measurement in order to have more calculations" do not help increasing the importance of the simulation in the engine development process, because for the development and validation of simulation tools, simulation engineers require extremely meticulous measured data of the highest possible precision including an extensive documentation of the prototype set up, otherwise measurements will become useless in many cases. Test engineers, on the other hand, require results that are available in a short while and frequently also measurements at a low level of accuracy are sufficient to them for an evaluation or for recognizing a relevant trend. Without a doubt, goals diverge greatly here. The target of the reduction of measurements in this case would probably result in reducing especially measured data which are useful for simulation engineers, as the main priority is that test engineers have to primarily observe their series production deadlines and development budgets.

2.2   Today's Repartition of the Resources in Engine Development

The analysis of an estimation of the currently typical repartition of test stand and computational resources (see Figure 2.4) clearly reveals that the investments for test stand activities are predominant. In particular "mechanical" investigations at the test stand (e.g. engine endurance run tests) still have a very high proportion of the total resources [10].

Accordingly, experimental "mechanical" verifications for ensuring the durability of the engine during the programmed life, despite the intense use of FEM and MBS simulations, are still extensive and very expensive. Without doubts this is a right and well consolidated practice, because otherwise, trying to save expense here could have fatal consequences with costs that would by far exceed the savings in engine development (e.g. recall campaigns). Even worse, long-term loss of product and company image has to be taken into account. The reason for this repartition is that the highest priority in any engine development is:

> *An engine must "keep",*
> *otherwise it cannot be produced or sold!*

Frequently in resource considerations, no differentiation is made between "application proportion" (engine maps application) and the actual combustion process and exchange process development. If these two aspects are separated it becomes clear that the expense incurred for the provision of data for the electronic control units (ECU) is markedly higher than the actual combustion process development. The reason is to be found in the second highest priority of any engine development.

> *An engine (vehicle) must "pass" the required exhaust gas tests,*
> *otherwise it may not be produced and sold!*

All other factors (fuel consumption, noise, etc.) are secondary to these two priorities. Short development periods can also entail that deficiencies in the combustion process development occasionally have to be accepted at first and then in a second step have to be eliminated with an improved application. Frequently this has been done with great success.

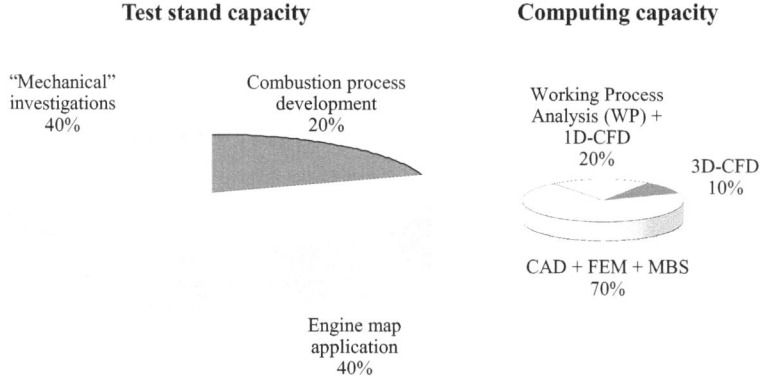

Test stand capacity

"Mechanical" investigations 40%

Combustion process development 20%

Engine map application 40%

Computing capacity

Working Process Analysis (WP) + 1D-CFD 20%

3D-CFD 10%

CAD + FEM + MBS 70%

Figure 2.4: *Current typical repartition of both test stand and computing capacities for development of internal combustion engines.*

Since test stand capacities alone are determinant for the final validation of an engine and also in the near future it cannot be expected that simulation tools will decisively support this task, it is evident that resources for computing capacities still represents a minor part of the total investment for engine development. Despite that, there is no doubt that the role of simulation tools does not have to be diminished; their benefit can be described with the following motto:

Thanks to simulation tools the engine of the future is not virtual
(i.e. investments for test benches will not be remarkably reduced);
it is a better one!

Focusing on the resources for computing capacities, 3D-CAD applications with MBS and FEM simulations clearly represent the most relevant, resource-consuming and established part of the virtual engine development process. Designing a new engine has always been "virtual work" from the very beginning of engine development. In former "pre-3D-CAD" times, however, the three-dimensional image of the engine existed only in the "engineer's head". Accordingly, coordination within design groups was difficult when complex designs were concerned.

Nowadays, 3D-CAD design permits not only the visualization of the engine but also to analyze the suitability for manufacturing, the assembly and in a certain extent the kinematic functionality. After the 3D-CAD design, MBS and FEM simulations allow to investigate the mechanical and thermal load of each part of the engine so that design refinements with improved solutions can be carried out. Without the reliable support of MBS and FEM simulations it would be unthinkable that even the first prototypes of a newly developed engine would have a good chance to withstand a 500 hours endurance run at the test bench.

In the last two decades, in order to assist the engine design process an increasing reliance has been set in the systematic use of simulation tools for the analysis of the operating cycle. By means of the real working-process analysis (WP), 1D- and 3D-CFD-simulations it is nowadays possible to gain, at least, a deep insight of the processes that govern engine performance and emissions.

Apart from FEM simulations, it is mostly the calculation result of a 3D-CFD simulation which is shown when a "virtual engine" has to be imaged. Here the detailed analysis of the reactive flow field in the domain of interest within the engine (usually the combustion chamber) is extremely impressive. Thanks to this peculiarity, e.g., any occurring phenomena (engine process) independently from its complexity can be numerically isolated and investigated, the geometry and motion of each part involved in the simulation is finely reproduced and can be easily varied for testing different designs with a high level of predictability. For these reasons the 3D-CFD-simulation should be the "silver bullet" of the engine design process and should not deserve such a limited figure of resources as reported in Figure 2.4.

The limitations in the application of the 3D-CFD-simulation have to be searched in the unsatisfied reliability and low level of maturity of some implemented models for the reproduction of some phenomena, i.e. a more detailed analysis here often means not necessarily a more correct one. Another, not less important drawback of the 3D-CFD-simulation (see Figure 2.5) is the required CPU-time that very often does not match the tight timetables in engine development.

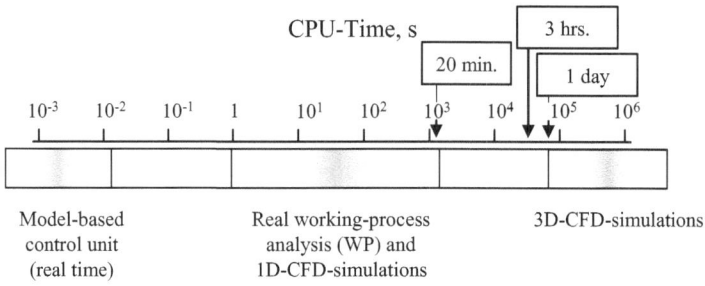

Figure 2.5: *Spectrum and CPU-time of the calculation/simulation tools for the analysis of the engine operating cycle.*

Looking at the required CPU-time of the various simulation approaches it becomes obvious that the real working-process analysis (WP) alone or in combination with the 1D-CFD-simulations represents a great potential here. A stand-alone real working-process analysis (WP) still takes about 10,000 more time than model-based control units (ECU) which have to calculate in real time during the engine operation - this also illustrates how great the restrictions are with the models implemented in these ECUs – but they can be up to 10 to 30 times faster than the required time for the measurement of a single operating point at the test bench. Accordingly, a greater number of parameter variations can be calculated in the same time period that can be measured. Moreover, this can be done at a remarkably lower cost, around the clock, without repeat measurements or test-stand maintenance. A stand-alone real working-process analysis (WP) of course has a quite limited predictability (e.g. investigations of different ignition timings on the efficiency of the working period), but in combination with a computational time-efficient calculation of the engine exchange-process, it becomes today's most powerful development tool.

Apart from structural calculations, real working-process (WP) and 1D-CFD-simulations nowadays are a self-evident part of any engine development. As early as during the concept, preliminary-investigation and design phase, the complete dimensioning of the intake and exhaust systems ensures that even the first test sample of the complete engine approaches the required specifications relative closely (e.g.: prediction of the volumetric efficiency to be expected under full load and fuel consumption calculations in official test cycles including the optimization of gear stepping).

2.3   Introduction to Engine Processes Modeling in the Simulation of the Operating Cycle

> *One of the principal objects of practical research...*
> *is to find the point of view from which*
> *the subject appears in its greatest simplicity [20].*
>
> *J.W. Gibbs*

The processes occurring in the working fluid of an internal combustion engine are numerous and very complex. Therefore an accurate numerical modeling of the processes is required. For this task, two basic approaches have been developed. As mentioned before in introducing the simulation tools these can be mainly categorized as thermodynamic or fluid dynamic in nature, depending on whether the implemented equations are based only on energy conservation (zero-dimensional) or on a full analysis of the fluid motion (multi-dimensional) [5,7,10,12].

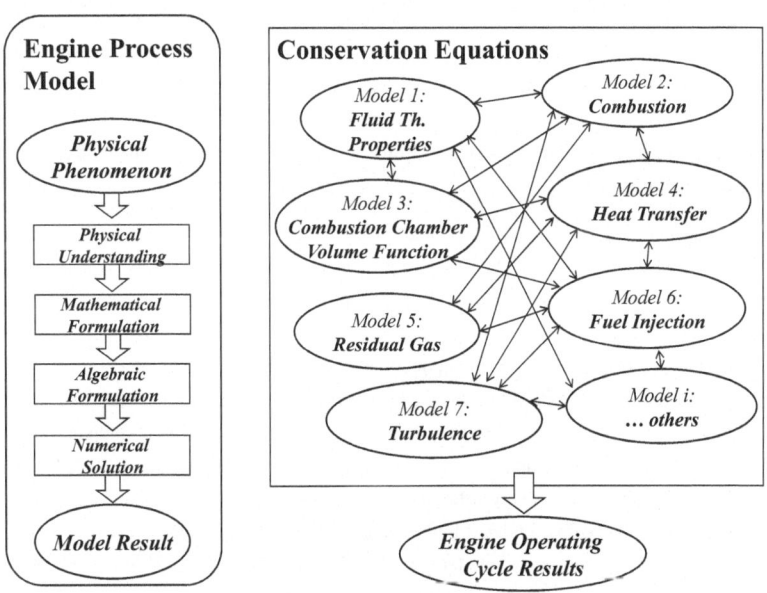

Figure 2.6: *Engine Process Modeling.*

An engine simulation tool independently from its approach is actually a collection of various engine process models (physical, chemical and thermodynamic phenomena or just control/actuation models like: injector model, spark ignition model that in a successive step generate a phenomenon). Here the conservation equations in the simulation tool (see Figure 2.6) are the "webs" for all the information transfers among the engine process models that are needed for calculating the final results.

Although the conservation equations that set the balance of the thermodynamic engine-processes are well known and evident like in any other thermodynamic system, the procedure towards a reliable and appropriate process modeling does not follow a unique way and still represents a most complicated and controversial task. Due to the complexity of engine processes, the insufficient understanding at fundamental level and very often still the limited computational resources, it is not possible to model engine processes that describe all important aspects starting from the basic governing equations alone. First of all, in order to govern the complexity, each modeled process must be limited to its relevant effects on engine behavior that have to be analyzed, then the formulations of the critical features of the processes have to be based on a keen combination of assumptions, approximations, phenomenological and eventually empirical relations. This procedure permits both to bridge gaps in our phenomena understanding and to lessen the computational time by reducing the number of required equations [5,7,10,12,21,22,23].

For any simulation approach, depending on the context, the formulations of engine processes that have stood the test of time show different level of sophistication. Each of them is able to predict with varying degrees of completeness, versatility and accuracy the predominant structure of the investigated process.

The formulation of engine processes is an active practice that continues to develop as soon as our understanding of the physics and chemistry of the phenomena expands and as soon as our ability to properly convert the process understanding first into a mathematical formulation, then into an algebraic formulation and finally into a numerical implementation, increases (see Figure 2.6). Any of these steps between the physical phenomenon and the resulting model is responsible for the general accuracy of the model. E.g. a particular emphasis only on the numerical discretization would for sure not lead to a more accurate analysis. Similarly any engine process model in the complex information exchange of the simulation tools is not more accurate than its weakest link. The weakest links are often not only represented by the mathematical formulation of the model or by the problem related to the numerical implementation but also by the accuracy of the input variables from other models. None model can provide reliable results until their inputs are reliable and, in case of a 3D-CFD-simulation, mesh independent. In conclusion each critical phenomenon should always be described by engine models at comparable levels of sophistication in order to establish a balance of complexity and details amongst the engine process models.

The development and validation steps of a 3D-CFD-model are a particular challenge. Due to its spatial and temporal resolution a 3D-CFD-model implemented into an existing 3D-CFD-code and applied to an engine mesh is able to locally reproduce an engine process (local implementation) but its results, if at all, can be measured and validated only globally (see Figure 2.7). For this reason, due to the absence of reliable local measurements of the investigated process on a real "unmodified" engine, the validation of the 3D-engine models still represents a critical and often "subjective" step.

A more simple validation way (validation with model tests) is the application of the 3D-CFD-model to a mesh reproducing a labor device (e.g. a pressure chamber) that allows a more accurate comparison with experimental measurements. Unfortunately the simulated and measured processes are, under labor conditions, far away e.g. from the real behavior of the process in the combustion chamber of an internal combustion engine.

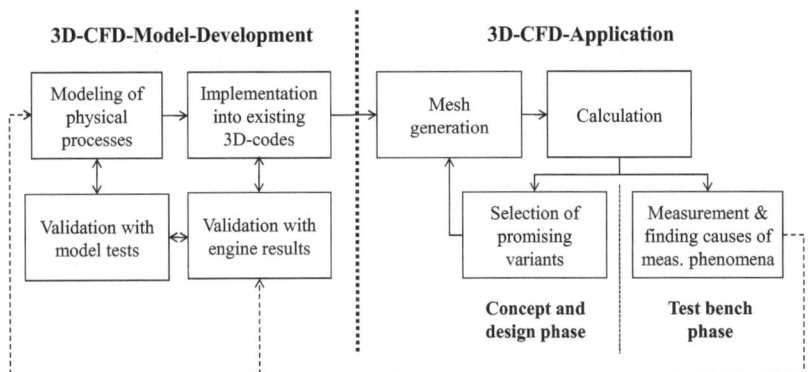

Figure 2.7: *Development, validation and application of engine process models for the 3D-CFD-simulation.*

The desire to improve the engine process modeling in the 3D-CFD-simulation is very high. First of all, this would allow to visualize and to investigate all relevant phenomena, also that ones that elude recording by measuring devices and then, after a more comprehensive evaluation of the engine behavior, it would allow a more efficient and rapid selection of promising variants. Especially in the near future the emphasis on process modeling will remarkably gain in importance; in particular, the necessity to explore new engine concepts towards a drastic reduction of CO_2 and pollutant emissions, respectively, will require robust simulation programs able to manage the rising complexity.

In the following chapters (3-6) a more detailed description of the engine simulation approaches is reported. It is not the aim of this work to be a compendium of engine modeling; hence it is more purposeful to try to understand the motivation, advantages and limitation of the different simulation approaches and their engine process models depending on both the application and the context for which the chosen simulation tool is asked to give information.

In the other chapters, as the core part of this work, an innovative fast-response 3D-CFD-tool called *"QuickSim"* is presented. This tool, developed during my activity at both the FKFS (Research Institute for Automotive and Internal Combustion Engines - Stuttgart) and the IVK-University of Stuttgart, introduced a new concept in the simulation of internal combustion engines that aims to increase the relevance and reliance of the 3D-CFD-simulation in the engine development process.

3

Engine Energy-Balance

Engine energy-balance and the real working-process analysis are the oldest and most validated approaches for the calculation of the engine operating cycles. Since the beginning they have accompanied the engine development processes. In the approach of the engine energy-balance, the combustion chamber, the entire engine or any other part of an internal combustion engine can be treated as an open thermodynamic system. Target of this approach is the analysis of the energy exchanges of the system through its boundaries with the surrounding system. In case of the entire engine the surrounding is represented by the environment.

3.1   Energy-Balance of the Combustion Chamber

The boundaries of the system are represented by the cylinder walls (head, piston, liner and valve heads), valve ports and piston crevice (see Figure 3.1). Since the combustion chamber is defined only by the volume V and area A_W, i.e. no geometrical details are taken into account, the analysis of the fluid motion is skipped [5,6,7,8,24,25]. Supported by the *mass conservation equation* (see Eq. 3.1 for a gasoline engine with port injection) the energy-balance based on the first law of thermodynamics (*energy conservation equation* - see Eq. 3.2) reports the most commonly operating parameters over the cycle such as the indicated work W_{Ind}, the fuel heat-release Q_B and the wall heat transfer Q_W through the combustion chamber. In addition, relevant parameters for the exchange process are also taken into account; these are the mass m_I and the thermal enthalpy H_I of the fluid through the intake valve, respectively, m_E and H_E through the exhaust valve and finally m_{BB} and H_{BB} as the blow-by across the piston rings.

$$m_I - m_E - m_{BB} = 0 \qquad (3.1)$$

$$Q_B - W_{Ind} - Q_W + H_I - H_E - H_{BB} = 0. \tag{3.2}$$

In these two fundamental equations of the energy-balance mass and energy flows are taken positive if they are in accordance with the scheme of Figure 3.2.

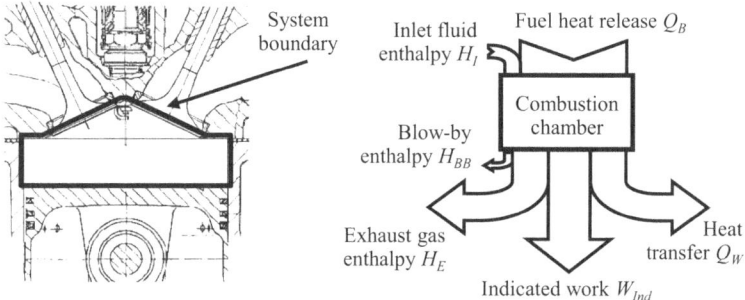

Figure 3.1: *The combustion chamber as an open thermodynamic system.*

Figure 3.2: *The energy balance of the combustion chamber (Sankey diagram).*

Based on the conservations equations the energy balance is principally a calculation procedure that permits to relate the energy terms of the system. Usually few of these terms (like Q_B, m_i and W_{Ind}) are mainly provided by measurements at the test bench so that the energy balance becomes the first approach that links an experimental investigation into a numerical analysis. Actually the energy balance is not a predictive tool, although in the past a collection of empiric models (black-box) have been developed. These models validated with numerous experimental investigations allow to estimate - depending on the engine speed, load and air/fuel ratio - the indicated work, the heat transfer and the exhaust enthalpy starting from the fuel energy content.

3.2   Energy-Balance of the Entire Engine

Assuming the entire engine as the thermodynamic system (see Figure 3.3 and Figure 3.4), the boundaries are represented by the intake, fuel and exhaust device, respectively, the cooling system and the environment. The conservation equations in this case are given by:

$$m_{I,Engine} - m_{E,Engine} = 0 \tag{3.3}$$

$$Q_B - W_{Eff} - Q_{Cooling} + H_{I,Engine} - H_{E,Engine} - Q_{Env} = 0. \tag{3.4}$$

The work W_{Eff} is the effective one at the brake of the test bench. Depending on the engine operating condition the cooling heat $Q_{Cooling}$ (water, oil and eventually EGR) principally collects the main part of the heat transfer from the combustion chamber Q_W, a part of the enthalpy flow of the exhaust gas H_E (wall heat transfer in the exhaust channels) and the friction work W_R;

$$W_R = W_{Ind} - W_{Eff} \tag{3.5}$$

The term Q_{Env} (energy that has not been removed by the engine cooling) represents the convective heat transfer of the exterior surface of the engine and the radiations to the environment. Here, according to the energy balance, the enthalpy flows of the intake $H_{I,Intake}$ and exhaust gas $H_{E,Engine}$ are calculated at the section of the engine pipes with the boundaries of the thermodynamic system.

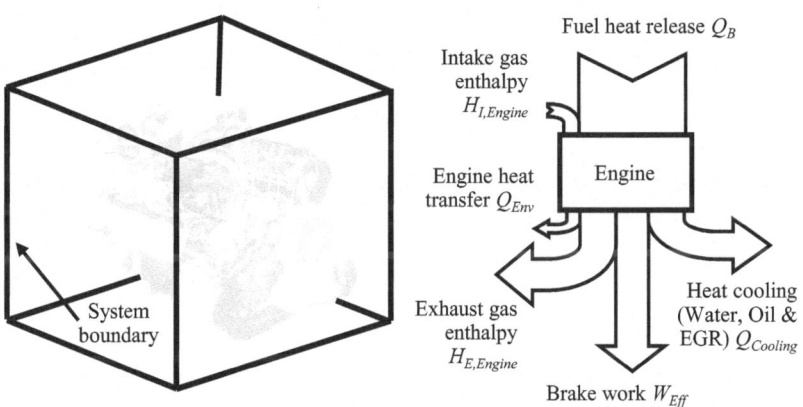

Figure 3.3: *The entire engine as an open thermodynamic system.*

Figure 3.4: *The energy balance of the entire engine (Sankey diagram).*

Similarly to the energy balance of the combustion chamber, also in this case, a collection of empiric models (black-box) has been developed in the past with the target to allow a plausible estimation of the repartitions of the energy terms.

3.3   The Role of Engine Energy-Balance in the Engine Development Process

The engine energy-balance does not provide deep insight of the behavior of the processes within the cylinder. Instead of this, it acts as the first irreplaceable stage towards an expressive thermodynamic characterization of the investigated engine over the full range of operating conditions (first link between experiments and numerical analysis). The most relevant results here, like torque, power, the hourly air and fuel consumption (i.e. the CO_2 emissions can be easily calculated), the thermodynamic efficiency (indicated efficiency) and the heat transfer are important also for a preliminary basis layout of the engine (dimensioning of the engine parts, the cooling system, etc.).

In engine simulating, among other things, the results of the engine energy-balance (thermodynamic evaluation) have great importance because they serve as reference values for the validation of more sophisticated and detailed simulation tools, i.e. the engine energy balance is the "supervisor" for other simulation tools.

> *Any simulation tool, independently from its complexity,*
> *has to pass the plausibility test with the energy balance,*
> *otherwise it cannot be used!*

This sentence has accompanied my work since the beginning.

4

Real Working-Process Analysis

The real working-process analysis is actually the direct evolution of the engine energy balance. The engine here is not a "thermodynamic black box" anymore, i.e. the thermodynamic cycle becomes the focus of this approach. Since the ideal p-V-diagram of an engine remarkably differs from the real one an accurate analysis with a reliable tool like the real working-process analysis is mandatory.

4.1 Introduction

The real working-process analysis (WP) as a time-resolved energy-balance of the combustion chamber permits a deeper insight of the progression of the main thermodynamic processes inside the cylinder during each phase of interest (see e.g. Figure 4.1 for the working period). Relevant for this analysis are the thermodynamic processes that directly affect the mass inside the combustion chamber and its energy balance (see Eq. 3.1 and 3.2).

The real working-process analysis is the engine simulation approach for the analysis of the operating cycle in which, until now, has been invested the greatest research efforts. This has been principally required by the necessity to find a way to "convert" time-dependent experimental data from the engine test bench –mainly represented by the pressure traces $p(\varphi)$ in the cylinder and in the manifolds, respectively – into a more comprehensive thermodynamic analysis of the operating cycle (see Figure 4.2). Hence the real working-process analysis is without doubt the most appropriate simulation tool for interfacing the engine test bench. Until the end of the eighties the CPU-time of computers was so high that such analyses were possible only in offline modus (post-processing).

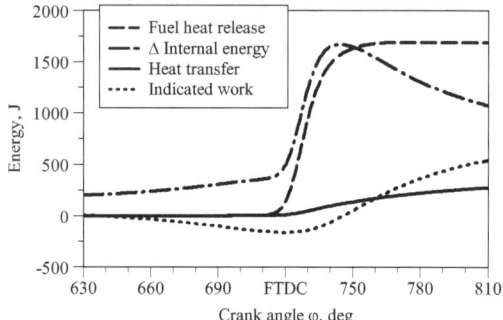

Figure 4.1: *The real working-process analysis (cumulative).*

Nowadays innovative real working-process codes are integrated in pressure indicating systems [7,12,26] and analyses of the working period are reliably performed online with experimental investigations at the test bench. Thanks to this approach a direct evaluation of the engine operating-conditions can be ensured.

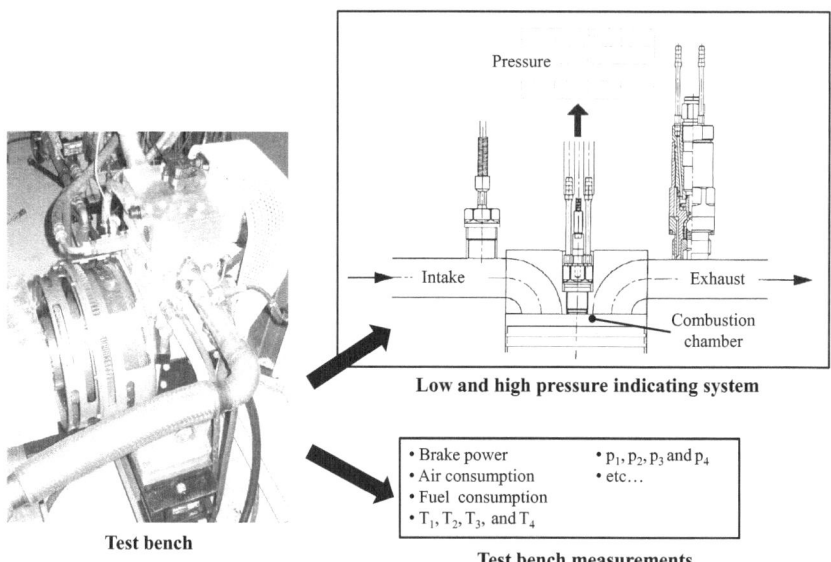

Figure 4.2: *Experimental investigations at the test bench.*

The real working-process analysis usually assumes uniformity in composition and state of the working fluid inside the cylinder at any time (one-zone-approach, see Figure 4.3), where T and R are respectively the temperature and the real gas constant.

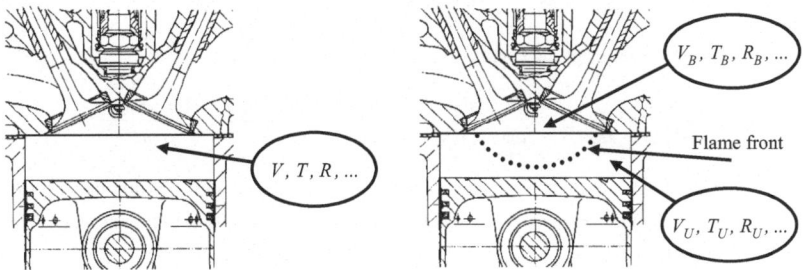

Figure 4.3: *One-zone-approach in the real working-process analysis.*   **Figure 4.4:** *Two-zones-approach in the real working-process analysis (SI-engine).*

During the combustion, temperature and composition change drastically in the burned zone. For this reason it is appropriate to divide the combustion chamber, at this time a closed thermodynamic system, into two interacting open thermodynamic systems: burned $(V_B, T_B, R_B, ...)$ and unburned zone $(V_U, T_U, R_U, ...)$ - (see Figure 4.4). In particular for SI-engines this is a common and convenient practice because burned and unburned zone are divided by a well defined small flame region. In case of more complex combustion processes (e.g. DI-engines with stratified mixture) it is often convenient to extend the partition of the combustion into more zones (n-zones approach), where the sum of the homogenous single ones better represents the real condition within the combustion chamber.

4.2   Fundamental Equations

The fundamental equations in the real working-process analysis are represented by the overall *conservation of mass and energy* in cylinder (Eq. 4.1 and Eq. 4.2 for the one-zone approach with algebraic sign convention as in Figure 3.2), where $U = mu$ is the internal energy of the working fluid. The total variation of internal energy over the entire cycle is $\Delta U = 0$ according to the engine energy balance (see Eq. 3.2). The two conservation equations, usually with the crank angle φ as the independent variable, set the balance of the thermodynamic engine-processes that take place in the combustion chamber and are the building blocks for the thermodynamic-based model:

$$\frac{dm}{d\varphi} = \frac{dm_I}{d\varphi} - \frac{dm_E}{d\varphi} - \frac{dm_{BB}}{d\varphi} \tag{4.1}$$

$$\frac{dU}{d\varphi} = \frac{dQ_B}{d\varphi} - \frac{dW_{Ind}}{d\varphi} - \frac{dQ_W}{d\varphi} + \frac{dH_I}{d\varphi} - \frac{dH_E}{d\varphi} - \frac{dH_{BB}}{d\varphi} \tag{4.2}$$

or by developing the terms of the energy conservation equation:

$$\frac{d(mu)}{d\varphi} = \frac{dQ_B}{d\varphi} - p \cdot \frac{dV}{d\varphi} - \frac{dQ_W}{d\varphi} + \frac{d(m_I h_I)}{d\varphi} - \frac{d(m_E h_E)}{d\varphi} - \frac{d(m_{BB} h_{BB})}{d\varphi}. \tag{4.3}$$

In case of more zones the mass and energy conservation equations must be formulated for each zone and also exchange terms for the interactions among the zones have to be introduced (more details in [5,7,26]).

4.3   Thermal State Equation of the Working Fluid

In addition to the conservation equations the well known thermal state equation of general formulation – *the perfect gas equation of state* - is used to relate temperature T, pressure p and density $\rho = m/V$ of the working fluid in the cylinder:

$$pV = mRT. \tag{4.4}$$

In the definition of the real gas constant R, considering the characteristic range of temperature and pressure inside internal combustion engines, it is not required to take real gas effects into account. A relationship for ideal gases is then used to link the specific internal energy u with the specific enthalpy h:

$$h = u + RT. \tag{4.5}$$

4.4   Engine Modeling (Engine-Specific Models)

The equations listed above, in case of an ideal gas, have general validity for any open thermodynamic system and are evident. Since the number of variables is much higher than the number of available equations, in order to close the equation system the thermo-physical properties of the working fluid and the terms that report the engine processes have to be defined.

As mentioned in Chapter 2.3 the procedure towards a reliable and appropriate process modeling does not follow a unique way and still represents the most complicated and controversial task.

The real working-process analysis mostly requires zero-dimensional formulations combined with additional adaptations that permit critical engine processes to be modeled. These formulations, usually, do not have general validity; therefore they are limited only to the modeling of engine processes (engine-specific models). These models describe the thermo-physical property of the working fluid and the main relevant engine processes that play a decisive role in both the engine energy-balance and the engine behavior (e.g. heat transfer, combustion, charge exchange process, pollutant emissions). In order to simplify matters, this chapter introduces only the models and their basic formulations that are required for the solution of the energy conservation equation during the working cycle (see Eq. 4.2).

4.4.1 Modeling of the Thermo-Physical Properties of the Working Fluid

The working fluid in the working-process analysis can be assumed as a mixture composed by three basic gases: air, fuel and exhaust gas (EGR and burned gas). The models that describe the thermo-physical properties of the fluid (real gas constant R and specific internal energy u) are almost always functions of temperature T, pressure p and the relative air/fuel ratio λ :

$$R = \frac{\Re}{\overline{M}} = f(T, p, \lambda) \tag{4.6}$$

$$u = \frac{U}{m} = f(T, p, \lambda) \tag{4.7}$$

where $\Re$ is the universal gas constant and $\overline{M}$ is the molar mass of the gas mixture.

An accurate and efficient modeling of the thermo-physical properties of the fluid is a complex task. The complexity increases in particular in the modeling of the exhaust gas produced by the combustion of rich mixtures due to the difficulties of determining the gas composition with satisfactory accuracy. In the last decades several mathematical formulations have been proposed. Some of them, the oldest ones, are exclusively based on an empirical approach [27,28]. More recent formulations for the exhaust gas are based on simplified reaction schemes of a reduced number of chemical species present in the burned mixture (usually ca. 9-14 species) [26,29-35]. The latter are of a more general formulation. The composition of the exhaust gas is calculated using the *condition of chemical equilibrium* at temperatures above about 1,700 K and a frozen composition below 1,700 K. These models are gaining in accuracy and also allow taking more complex effects (e.g. dissociation of CO_2, H_2O and N_2) into account.

4.4.2   Modeling of the Wall Heat-Transfer

At the historical beginning of the real working-process analysis implementations, the wall heat-transfer process has been calculated by empirical correlations [27,28]. Since the accuracy of these formulations was generally unsatisfactory, starting from the end of the sixties, heat-transfer models based on a more promising approach, the *phenomenological approach*, have been developed and implemented in the real working-process analysis. In the phenomenological approach the instantaneous spatially-averaged convective heat-transfer coefficient $\alpha(\varphi)$ is modeled [26,36-41], so that, in case of the one-zone approach, the heat transfer through the combustion chamber surface $dQ_W/d\varphi$ in contact with both the unburned and the burned gas zones is given by:

$$\frac{dQ_W}{d\varphi} = \alpha(\varphi)\, A_W(\varphi) \cdot \left[T(\varphi) - T_W\right] \frac{dt}{d\varphi} \qquad (4.8)$$

where $A_W(\varphi)$ and T_W are the instantaneous combustion chamber area and temperature, respectively. Due to the thermal inertia of the engine parts in comparison to the engine processes, the wall temperature of the combustion chamber surfaces is assumed constant during the operating cycle. The modeling of the convective heat-transfer coefficient $\alpha(\varphi)$ is nowadays based on the Re-Nu-correlation (dimensional analysis) in which the assumption of the Nusselt, Reynolds and Prandtl number relationship follows basically that one found for turbulent flow in pipes. Few dedicated additional adaptations make the correlation more suitable for simulation of internal combustion engines (engine-specific modeling). The most widely used correlations are, chronologically reported, from Woschni [39], Hohenberg [40] and Bargende [26] (more details in Chapter 10.2.3).

4.4.3   Modeling of the Combustion Process

Numerous models for the calculation of the combustion process have been proposed. More correctly these models should be called "heat-release models" because the analysis of the chemical reaction mechanisms is not their aim. Therefore, what is absolutely sufficient for a thermodynamic approach like the real working-process analysis, the calculation is limited to the heat release progression. These models can be generally divided into three main categories:

Empirical Models: The combustion process in terms of the progression of the burned mass fraction over the crank angle is described using an empiric function tuned by a limited number of coefficients. In this approach the analysis of the involved physical phenomena is rigorously skipped, i.e. by varying the engine operating conditions or the combustion chamber design no predictability has to be expected.

Quasi-Dimensional Models: "Quasi-dimensional" or "phenomenological" combustion models dramatically reduce the physical and chemical reactions of the real combustion to a simple physical formulation, trying to identify and set as inputs the main variables able to influence the combustion process. In this way a quasi-dimensional combustion model can simulate the changing burn rate by varying the engine operating conditions, injection strategies, etc. Such models also take into account both few geometrical parameters of the combustion chamber and the fuel jet shape. Therefore a certain predictability can be ensured. In comparison to the 3D-CFD-simulation the geometrical approach of these models is much simpler (for that reason these models are called "quasi-dimensional") but on the other hand the computing time can be drastically reduced (1 second for the simulation of one operating cycle instead of many hours required by a 3D-CFD-simulation).

3D-CFD-Models: The full analysis of the flow field allows a detailed calculation of the progression of the flame propagation within the combustion chamber. Since all relevant geometrical details are also taken into account (see Chapters 6 and 9) these models ensure a very high predictability. They cannot be implemented into the real working-process analysis as long as the combustion chamber is not finely discretized (prerequisite of the 3D-CFD-simulation where also in case of a fast response tool like *QuickSim* usually about 30,000 cells are required for the discretization of combustion chamber).

Below a brief description of the empirical and quasi-dimensional approach is given. The three-dimensional approach will be treated in Chapters 6, 8, 9 and 10.

4.4.3.1 Empirical Models

Empirical models have been principally formulated during the time (the sixties) of engine development of quite simple SI-engines with moderate specific power output, no variable valve strategies, no direct injection (homogeneous mixture), etc. Here the engine map is represented by a simple speed and load field, so that the empirical combustion models can be easily calibrated and the lack in predictability is not a serious matter.

Combustion Process in SI-Engines with Homogeneous Mixture

During the combustion, starting with the inflammation at the spark plug at the ignition point (IP) a flame, first laminar and then turbulent, propagates inside the combustion chamber within an essentially premixed mixture of fuel, air and residual gas (see Figure 4.4). At the flame front numerous exothermic chemical reactions (mainly fuel oxidation) are involved. These cause a rapid increase of the temperature of the burned gas followed by a raise of the pressure in the cylinder. The flame extinguishes, when it, reached a small distance from the walls, due to wall heat-transfer and a lower diffusive process that supplies the flame, the temperature in the reaction zone drastically decreases to a critical value at which the combustion processes cannot be supported anymore. At the end of the combustion φ_C always a small amount of fuel still has not been involved in any oxidation process (*imperfect combustion*). This unburned fuel, which is a source of pollution (UHC), has to be oxidized in the catalyst after leaving the combustion chamber.

The total amount of energy Q_B that is released by the fuel can be defined by:

$$Q_B = \eta_{HR} \cdot \eta_C \cdot m_F \cdot h_{LHV} \tag{4.9}$$

where m_F is the fuel mass trapped in the cylinder, h_{LHV} the fuel's lower heating value, η_{HR} the combustion conversion efficiency due to *incomplete combustion or incomplete oxidation* and η_C the combustion efficiency caused by *imperfect combustion*.

$$\lambda \geq 1 \quad \Rightarrow \quad \eta_{HR} = 1 \tag{4.10}$$

$$\lambda < 1 \quad \Rightarrow \quad \eta_{HR} = 1.3733 \cdot \lambda - 0.3733 . \tag{4.11}$$

Under lean operating conditions the amount of incomplete combustion products is small so that they can be neglected $\eta_{HR} = 1$ (complete fuel oxidation at the flame front). Under fuel rich operating conditions these amounts become more substantial since the oxidation of the fuel is incomplete. In this case, because a fraction of the chemical energy stored in the fuel cannot be fully released to the burned gas it is necessary to define a relation for the combustion conversion efficiency η_{HR}. A simple formulation proposed by Vogt is reported in Eq. 4.11 [42]. Also more comprehensive formulations have been proposed during the years [12,31,32,35].

During the combustion, in case of homogeneous mixture, the mass of burned fuel $m_{F\_B}(\varphi)$ at a crank angle φ is proportional to the mass in the burned zone m_B so that it is convenient to introduce the burned mass fraction variable w_B which describes the ratio of mass of the working fluid "caught" by the flame development to the total mass in the cylinder m (see Eq. 4.12).

$$w_B = \frac{m_{F\_B}(\varphi)}{m_F} = \frac{m_B}{m} \leq \eta_C. \qquad (4.12)$$

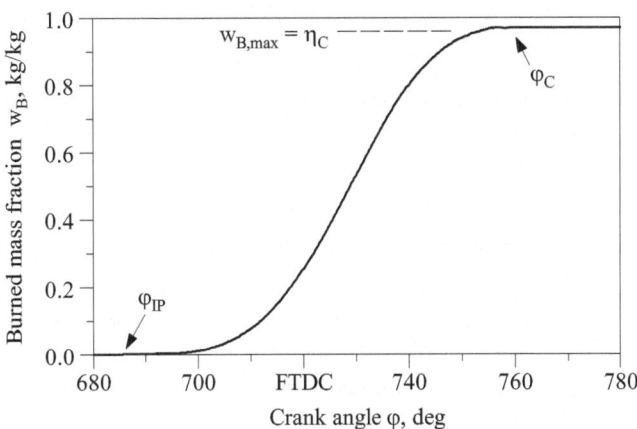

Figure 4.5: *Characteristic cumulative combustion profile (SI-engines).*

Since the combustion process is *imperfect* the burned mass fraction variable has to be limited by the combustion efficiency η_C. The fuel heat-release rate $dQ_B/d\varphi$ as a function of the burned mass fraction profile can then be described by Eq. 4.13.

$$\frac{dQ_B}{d\varphi} = \frac{dw_B}{d\varphi} \cdot \eta_{HR} \cdot m_F \cdot h_{LHV} \quad ; \quad w_B \leq \eta_C. \qquad (4.13)$$

In SI-engines with homogeneous mixture, the burned mass fraction profile *(cumulative combustion profile)* as a function of crank angle $w_B = f(\varphi)$, independent on engine load, always has a characteristic S-shape (see Figure 4.5). The combustion profile, even for simple internal combustion geometries, is influenced by countless factors acting locally at the flame front. In particular, among others, the local gas mixture composition (air/fuel ratio and concentration of residual gas), temperature, pressure and fluid motion (especially turbulence) have great influence on the flame development. In the real working-process analysis these local variables cannot be detected, i.e. there is no possibility in a zero-dimensional approach using empirical models to predict with accuracy the flame propagation law and then the combustion profile w_B.

In order to skip all the complexity related to the combustion process in this approach, functional forms are often used for describing the characteristic S-shape of the cumulative combustion profile $w_B = f(\varphi)$. The most known and applied function form is represented by the Wiebe function [5,7,26,43]:

$$w_B = \eta_C \left\{ 1 - \exp\left[-a \cdot \left(\frac{\varphi - \varphi_{IP}}{\varphi_C - \varphi_{IP}} \right)^{m+1} \right] \right\} \tag{4.14}$$

where a and m are adjustable coefficients for the calibration first and then for modeling the shape of the combustion profile. From the derivation of $w_B = f(\varphi)$ it is then possible to define the heat release rate:

$$\frac{dQ_B}{d\varphi} = \eta_{HT} \cdot \eta_C \cdot m_F \cdot h_{LHV} \cdot \left\{ a(m+1) \left(\frac{\varphi - \varphi_{IP}}{\varphi_C - \varphi_{IP}} \right)^m \exp\left[-a \cdot \left(\frac{\varphi - \varphi_{IP}}{\varphi_C - \varphi_{IP}} \right)^{m+1} \right] \right\}. \tag{4.15}$$

In Figure 4.6 the heat release rates of an SI-engine equipped with inlet-phasing for two different positions of IVO are reported [18]. Here it can be seen how the heat release rate profile is sensitive by varying the engine setting conditions.

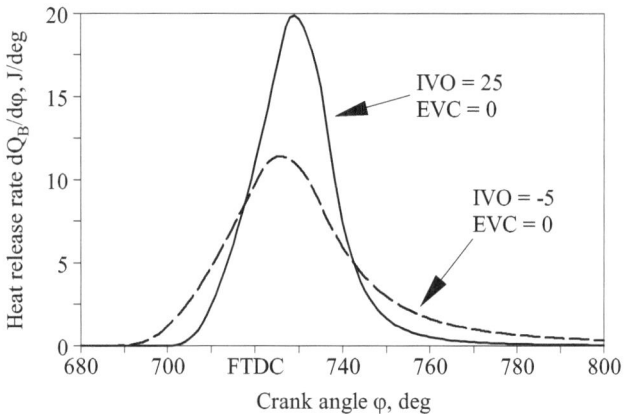

Figure 4.6: *Characteristic heat release profiles*
(homogenous SI-engines with intake valve phasing – 1,600 rpm – 3 bar imep).

In order to perform a reliable real working-process analysis over the whole engine map it is then necessary to properly set these coefficients for each investigated operating condition.

How to Master more Complex Combustion Processes ?

As introduced in Chapter 1 SI-engines with homogenous mixture are becoming an obsolescent concept. A new generation of engines like turbocharged direct-injection SI-engines (DI-engines) or innovative common-rail diesel engines and, e.g., future HCCI-concepts are characterized by much more complex combustion processes. In these cases the combustion progressions (see Figures 4.7 and 4.8) do not have a simple shape anymore and they are influenced by a greater number of additional factors (e.g. injection and turbo-charging strategy) than only engine speed and load [5,7,12-15,17,44].

In DI spark-ignition engines as a result of the injection directly in the combustion chamber during the compression stroke, it is possible to create a mixture that is ignitable only locally at the spark plug. This distinguishes them from conventional spark-ignition engines with homogeneous mixture in the combustion chamber. In an ideal case, the stratified charge DI spark-ignition engine can intake the maximum possible air quantity in partial load operation. Due to this de-throttling, the efficiency of the total process is significantly increased. Here, the entire fuel-air mixture in the combustion chamber is partly very lean. For stratified charge spark-ignition engines, special injection processes must be used to ensure that the required ignitable mixture is present at the spark plug. In addition also valve timing strategies, turbo-charging control or tumble-flaps are able to influence the charge motion and the EGR concentration within the combustion chamber, which remarkably influence the combustion process.

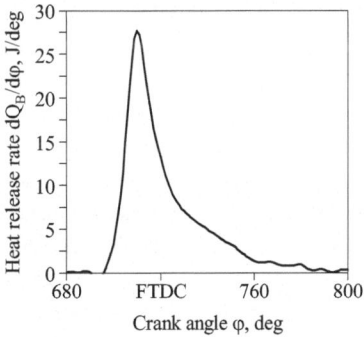

Figure 4.7: Characteristic heat release profile (direct injected SI-engine with stratified mixture – 3 injections strategy)

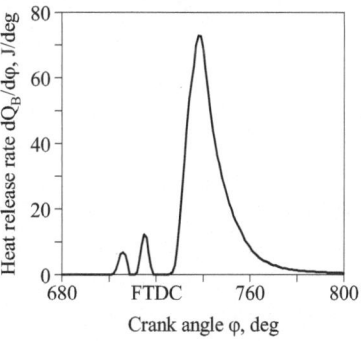

Figure 4.8: Characteristic heat release profile (diesel-engine with multi-injection strategy)

In modern diesel engines with common-rail injection systems the mixture formation and successively the combustion process can be freely modified. The injection system allows any variation of the injection pressure (responsible for the jet penetration), number of injections in an operating cycle (pilot, main and post-injection, etc.) and injection timing of each injection. In addition charge motion and composition can be varied by the control of the turbo-charging, the EGR-valve and the swirl-flap, which introduces additional influencing factors to the combustion process.

For these applications it is clear that empirical combustion models need additional adjustable parameters in order to establish a more comprehensive function able to reproduce more complex combustion profiles [45]. A greater number of adjustable coefficients surely make these formulations more flexible but on the other hand it exponentially complicates the calibration processes, so that at the end the process becomes very difficult to handle. For these reasons in case of complex combustion processes it makes sense to implement more innovative and recent models based on the quasi-dimensional approach in the real working-process analysis.

4.4.3.2 Quasi-dimensional Models

Due to simplicity a brief description of a quasi-dimensional combustion models is limited to SI-engines with a nearly homogenous mixture formation (no stratified mixture). More details about more complex combustion processes are reported in the literature [13,14,15]. Quasi-dimensional models are based on a simple physical and chemical prediction of the flame propagation within the combustion chamber.

This approach assumes flame propagation with the absolute velocity u_f as a hemisphere, which extends with a speed normal to its surface (see Figure 4.9). The combustion chamber is divided into three parts: unburned zone, burned zone and the flame front which separates the two zones. Here the flame front, because of the very small volume, can be assumed as thermodynamically irrelevant, i.e. it does not appear separately in the conservation equations and can be added to the unburned zone. Consequently the model corresponds to a two-zone combustion model. The combustion chamber is modeled as a disc with the same volume of the original geometry over the time. This assumption simplifies the approach and shortens the calculation time while still providing good results. A possible occurring error can be easily compensated by an overall adjustment of the model.

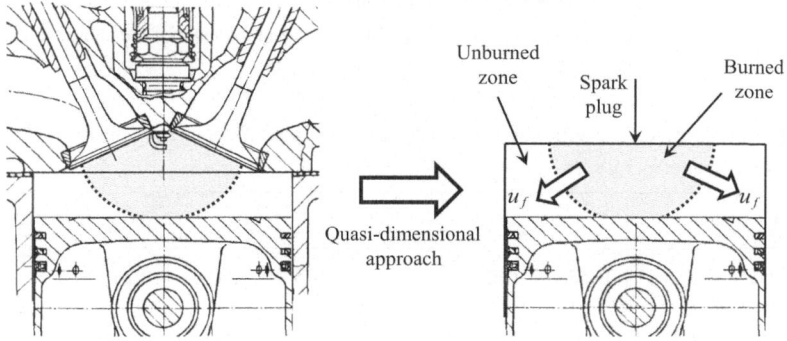

Figure 4.9: *Quasi-dimensional description of flame propagation (combustion model).*

Considering, conveniently for the modeling, the relative speed of the flame front S_T penetrating the unburned zone, S_T is assumed as the sum of the laminar flame speed S_L and an isotropic turbulence speed term u_T:

$$S_T = S_L + u_T .$$
(4.16)

The velocity S_T represents the relative velocity of the flame front to the unburned zone and remarkably differs from the absolute flame speed u_f that takes into account also the volume dilatation of the burned zone due to the decreasing density during the heat release ($T_B \cong 3 \cdot T_U$). The required laminar flame speed S_L of a fuel-air mixture can be provided, e.g., by the Guelder's correlation (Eq. 4.17) [46], for which a selection of fuel-specific coefficients is listed in Table 4.1. This correlation allows the modeling of the laminar flame speed S_L for different fuels under the following thermodynamic and chemical conditions: lambda λ, temperature in the unburned zone T_U, cylinder pressure p and mole fraction of the residual gas x_{EGR_U}:

$$S_L(\lambda, p, T_U, x_{EGR\_U}) = Z \cdot W \cdot \left(\frac{1}{\lambda}\right)^{\eta} \cdot \exp\left[-\varsigma \cdot \left(\frac{1}{\lambda} - \Phi_m\right)^2\right] \cdot$$
$$\cdot \left(\frac{T_U}{T_0}\right)^{\alpha} \cdot \left(\frac{p}{p_0}\right)^{\beta} \cdot \left(1 - F_{EGR} \cdot x_{EGR\_U}\right)$$
(4.17)

with:

$$x_{EGR\_U} = \frac{n_{EGR\_U}}{n_U} = w_{EGR\_U} \cdot \frac{M_U}{M_{EGR}}$$
(4.18)

where n_{EGR} and M_{EGR} are the number of moles and the molar mass of the residual gas in the unburned zone and n_U and M_U the number of moles and the molar mass of the unburned gas.

Table 4.1: Selection of coefficients for the laminar flame speed S_L of different fuels (Guelder's correlation).

Fuel	Z	W	ς	η	Φm	α	β	F_{EGR}
Isooctane	1.0	0.4658	-0.326	4.48	1.075	1.56	-0.26	2.1
Ethanol	1.0	0.465	0.25	6.34	1.075	1.75	-0.24	2.1
Methane	1.0	0.422	0.15	5.18	1.075	2.0	-0.5	2.5

Consequently the unburned mass brought into the flame front by the velocity S_T is:

$$\frac{dm_U}{dt} = \rho_U \cdot A_f \cdot S_T \tag{4.19}$$

where A_f is the area of the flame front and ρ_u the averaged density in the unburned zone. From the unburned mass rate the heat release during the combustion process can be related as follows [12]:

$$\frac{dm_U}{dt} = -\frac{dm_B}{dt} = \frac{dQ_B}{d\varphi} \frac{1}{\eta_{HR} \cdot h_{LHV} \cdot w_{F,U}} \frac{d\varphi}{dt} = \frac{m_f}{\tau_l} \tag{4.20}$$

$$w_{F\_U} = \frac{m_{F\_U}}{m_U} \tag{4.21}$$

where m_f is the mass of the flame front and w_{F_U} the fresh fuel mass fraction in the unburned zone. For a laminar combustion the characteristic burn-up time τ_l (see Eq. 4.22) represents the time required by the flame to pass a turbulence eddy with the size of the Taylor length l_{Taylor}:

$$\tau_l = \frac{l_{Taylor}}{S_L} . \tag{4.22}$$

Depending on the global length scale l_g (this scale does not have to be confused with the integral length scale l_i of the turbulence that describes the dimensions of the biggest turbulence eddies), the turbulence speed term u_T and the turbulent dynamic viscosity μ_T the Taylor length l_{Taylor} can be determined as follows:

$$l_{Taylor} = \sqrt{\chi_{Taylor} \frac{\mu_T \cdot l_g}{\rho_U \cdot u_T}}$$

(4.23)

where the global length scale l_g represents the characteristic dimension of the combustion chamber as a function of its volume V at any crank angle:

$$l_g = \left(\frac{6 \cdot V}{\pi} \right)^{\frac{1}{3}}$$

(4.24)

According to literature [5,12] the factor χ_{Taylor} is assumed to be 15. More details about turbulence and length scales can be found in Chapter 6.2.1.3.

Turbulence Modeling in the Quasi-dimensional Approach

In order to close the system of equations presented in the quasi-dimensional approach a formulation for the isotropic turbulence speed term u_T is required. This is for sure the most challenging part of this approach because, as well known, a detailed analysis of the flow field within the combustion chamber, at this degree of discretization it is not possible.

Starting from the assumption that turbulence is mainly influenced by both the tumble generation during the intake stroke followed by its breaking up during the compression stroke and squish effects when the piston approaches FTDC (for engines with piston bowl), during the years, several pragmatic and simplified approaches for zero- or quasi-dimensional turbulence modeling, respectively, have been presented. A detailed presentation of these isotropic $(\bar{k} - \bar{\varepsilon})_{0D}$-turbulence-models is extensively reported in the literature [7,12,26].

4.5 Two Approaches in the Calculation of the Real Working-Process

Focusing on the working period of the operating cycle, i.e. when the valves are closed, and neglecting blow-by flows due to simplicity, the energy conservation equation becomes:

$$\frac{dU}{d\varphi} = \frac{dQ_B}{d\varphi} - p\frac{dV}{d\varphi} - \frac{dQ_W}{d\varphi} \ . \tag{4.25}$$

As mentioned before (see Chapter 4.4.3), it is well recognized that the combustion profile cannot be zero-dimensionally predicted by solely solving the system equation of the real working-process analysis and also in case of quasi-dimensional combustion models an initial calibration is mandatory. In order to close the equation system (one thermodynamic degree of freedom) an "external" input as assumption is required.

Here two opposite approaches have been established:

- *pressure profile calculation* (combustion profile supply – simulation approach)

- *combustion profile calculation* (pressure profile calculation – experimental approach).

4.5.1   Pressure Profile Calculation - Combustion Profile Supply

In this approach in order to calculate the pressure profile in the cylinder $p(\varphi)$ the fuel heat-release rate $dQ_B/d\varphi$ has to be either known or assumed. From the energy conservation equation (Eq. 4.25) the pressure becomes:

$$p = \frac{\dfrac{dQ_B}{d\varphi} - \dfrac{dQ_W}{d\varphi} - \dfrac{dU}{d\varphi}}{\dfrac{dV}{d\varphi}} \ . \tag{4.26}$$

The derivation of the assumed combustion profile $w_B = f(\varphi)$, in case of homogenous mixture, then provides the required fuel heat-release rate $dQ_B/d\varphi$ so that the pressure profile in the cylinder $p(\varphi)$ can be calculated.

4.5.2   Combustion Profile Calculation - Pressure Profile Supply

In this approach, in order to calculate the combustion profile $w_B(\varphi)$, the pressure profile in the cylinder $p(\varphi)$ has to be either known or assumed. The energy conservation equation (Eq. 4.25) assuming a homogeneous mixture formation becomes:

$$\frac{dQ_B}{d\varphi} = \frac{dU}{d\varphi} + p\frac{dV}{d\varphi} + \frac{dQ_W}{d\varphi} \tag{4.27}$$

so that in case of a homogenous mixture:

$$\frac{dw_B}{d\varphi} = \frac{\dfrac{dU}{d\varphi} + \dfrac{dQ_W}{d\varphi} + p\dfrac{dV}{d\varphi}}{\eta_{HT} \cdot m_F \cdot h_{LHV}} . \tag{4.28}$$

In this approach it is important to verify that the maximum reached by the burned mass-fraction $w_{B,\max}$ (see Figure 4.5) assumes a value equal to a plausible combustion efficiency η_C *(imperfect combustion)*.

The combustion profile calculation is the most traditional approach in the calculation of the real working-process. As mentioned at the beginning of this chapter, nowadays complex real working-process codes are integrated in pressure indicating systems [7,12,21,22] and detailed analyses of the working period are reliably performed online with experimental investigations at the test bench that among other things provide the required pressure profile $p(\varphi)$ in the combustion chamber.

4.6   The Role of Real Working-Process Analysis in the Engine Development Process

As already mentioned in this chapter the real working-process analysis is the simulation tool for the investigation of the engine operating cycle in which, over decades, the most development efforts have been invested. At the moment no other simulation tool is able to interface experimental investigations at the test bench more efficiently and in a comprehensive matter then the real working-process analysis. In many applications this simulation tool is absolutely inseparable from experimental investigation, i.e. in these cases it acts as the extension of the measurement devices at the test bench allowing, as if a direct measurement were possible, an accurate analysis of the thermodynamic processes (wall heat-transfer, combustion, etc.) inside the cylinder during each phase of interest.

The processes in an internal combustion engine are extremely complex and, in particular at the discretization degree of the real working-process analysis, they cannot be solved starting from basic equations. This has required the development of more practicable and purposeful approaches for engine process modeling. In these approaches, among other things, a combination of phenomenological and empiric laws based on a zero-dimensional and eventually quasi-dimensional treatment of the thermodynamic system has permitted, after many years of modeling improvements often based on experimental tunings, to reach a very good accuracy and reliance in engine operating cycle analysis. The analysis requires low CPU-time and also investigations on very complex engines (DI-engines, diesel engines and also HCCI engines) can be performed by continuous increasing of result accuracy.

A powerful and important capability of the real working-process analysis is represented by the possibility to extend the modeling and analyses from the engine processes directly involved in the conservation equations (mass and energy balance) to other ones of interest, like exhaust gas emissions (NO_X, UHC, soot, etc.), knocking, mixture ignition, etc., that are today mandatory for a successful engine development process.

The real working-process analysis in combination with the 1D-CFD-simulation ("internal coupling" between WP and 1D-CFD) allows to extend the analysis from the combustion chamber as an open thermodynamic system to the whole engine, where the flow in the exhaust and intake system, respectively, can be properly one-dimensionally investigated up to a certain design complexity. Here the "internal coupling WP-1D-CFD" permits to increase the predictability of these simulation tools at least regarding the exchange process. Modern quasi-dimensional combustion models introduce, for any kind of engine, also a promising predictability in the calculation of the heat release based on a "coarse" analysis of the fluid motion within the combustion chamber. From a global point of view, these tools are getting more and more relevant and decisive in supporting the development of today's and future engines, respectively, where the number of parameters that can be varied has been and will be continuously increasing.

The following points are a selection of the main investigation topics using the real working-process analysis:

- Analysis of the engine operating cycle in all relevant thermodynamic terms (piston work, variation of internal energy, fuel heat-release, wall heat-transfer, direct fuel-injection and enthalpy fluxes through the valves) using pressure trace signals or combustion profiles

- Volumetric efficiency and residual gas calculation at IVC

- Starting from an accurate calibration of the models, modern real working-process analyses are able to ensure a good predictability by varying valve timings and injection strategies within a certain range of variation

- Extension to non-thermodynamic relevant models for the prediction of, e.g., exhaust gas emissions and knocking effects.

The limitations of the real working-process analysis have to be found mainly in the capability to analyze the influence of design details (e.g. a slight changing design of the piston surface or of the intake channel) on the fluid motion and all other processes connected, or generally in all applications where a detailed analysis of the flow field (e.g. mixture and residual gas distribution) is mandatory. For these tasks only the 3D-CFD-simulation is able to give answers.

As it will be discussed in the next chapters (6, 7, 8, 9 and 10) the experience in the real working-process analysis can be very helpful also in the 3D-CFD-simulation. Local 3D-CFD-engine models can remarkably take profit from the "well-tried" approaches in the WP; actually the 3D-CFD-simulation can be seen as the ideal extrapolation of the WP-analysis with a great number of zones n in the system ($n > 30,000$ zones for the combustion chamber). In addition an "internal coupling WP-3D-CFD" allows a reliable energetic control of the progression of the engine processes during the operating cycle calculated by the 3D-CFD-simulation and an effective comparison of these results with other simulation tools and experimental data.

5

One-Dimensional Simulation (1D-CFD-Simulation)

The 1D-CFD-simulation can be seen as an extension of the cylinder real working-process analysis. The latter provides mainly the calculation of the working period of each cylinder (i.e. when the valves are closed) while the 1D-CFD-simulation analyzes the flow field in the whole intake and exhaust track, respectively. This approach can ensure an accurate calculation of the engine exchange-process, so that e.g. reliable initial conditions for the real working-process analysis at intake valve close (IVC) can be provided.

5.1 Introduction

In the past the engine exchange-process analysis was limited to a simple calculation of only the flow through the intake and exhaust valves, respectively. In this case the mass flow through a poppet valve is usually described by the equation for compressible flows through a flow restriction. This formulation is derived from the one-dimensional isentropic flow analysis, where real gas flow effects are included by means of experimentally determined discharge coefficients. The region of the channel near the valves is treated as an open thermodynamic system of fixed volume which contains gas at both uniform state and uniform pressure. In order to close the problem it is necessary to determine the pressure profiles in the region near the valves of the inlet and exhaust channels, respectively. These pressure profiles, which have high influence on the engine behavior, are the result of complex interactions between the combustion chamber and the intake and the exhaust system, i.e. they cannot be easily estimated. For these reason the determination of the pressure profiles in the channels is based on experimental measurement at the test bench (see Figure 4.2), so that the real working-process analysis has most of the required information for a reliable calculation of the engine operating cycle [7].

A more detailed approach is represented by *filling and emptying models*. In this approach the intake and the exhaust system are represented by a coarse subdivision into finite volumes, usually only one finite volume for each manifold [47]. Each volume is then treated as a control volume - open thermodynamic system of fixed volume – which contains gas at both uniform state and uniform pressure. The mass of gas in the volumes as a function of time is calculated then by the continuity equation and the information concerning the pressure variations associated with momentum variations are not provided. This kind of models are a simple approximation of the reality. They allow acceptable accuracy of the results only for both simple geometrical structures of the intake and the exhaust system and for low amplitude of the pressure waves in the manifolds.

Finally with the development of 1D-CFD-tools, briefly described in this chapter, the analysis of the exchange process has became more comprehensive.

5.2   Engine Layout and Conservation Equations

In one-dimensional exchange-process models (see Figure 5.1), the whole intake and the exhaust system are discretized into many volumes (flow components). The pipes are generally divided into many segments while the junctions (flow-splits) are represented by a single one [48].

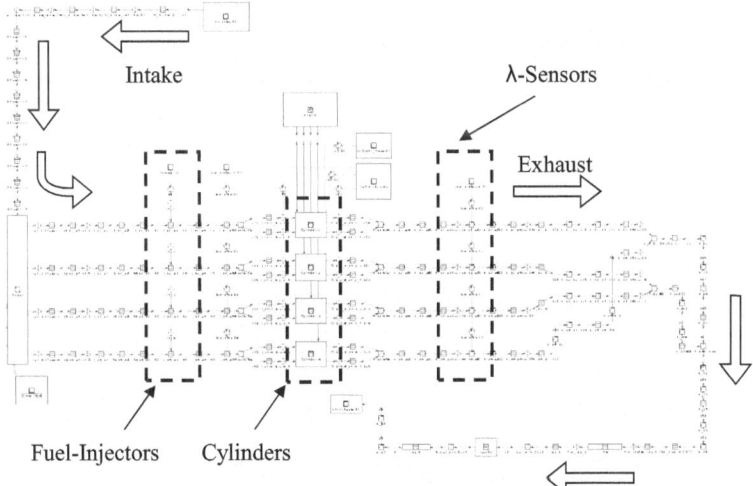

Figure 5.1: 1D-CFD-simulation: model layout.

Each flow component is referred to one template in order to accommodate a variety of geometries – pipes with either constant or conical diameter, flow-splits, throttles, valves, pressure loss connections, orifices, etc. This means any flow component can be geometrically defined by few parameters. Actually this type of discretization converts complex geometries into a flow network called "staggered grid".

Devices like turbo-chargers, fuel injectors, lambda sensors, etc., can be integrated into the flow network by means of modules (eventually linked to characteristic operation maps and the ECU-engine-tool). These modules aim to correctly reproduce the interactions of these devices with the flow in particular during a transient simulation.

The solution of the flow over the whole network involves the simultaneous simulation of the continuity, momentum and energy equations in each control volume. These equations are solved one-dimensionally, which means the all quantities are averages across the flow direction. Normally this approach allows a good accuracy of the results and an acceptable CPU-time (few minutes for each operating condition). However, the flow-network requires a calibration with experimental data at some operating conditions.

5.3   The Role of the 1D-CFD-Simulation in the Engine Development Process

As discussed in the previous chapters the 1D-CFD-simulation in combination with the real working-process analysis (WP) for the cylinders is a determinant tool for a successful engine development process. In particular the relative low CPU-time and the capability of a comprehensive analysis have remarkably increased the popularity of the 1D-CFD-simulation among engineers in the last years. Here a selection of few important development tasks shows how wide is its range of applicability.

- Intake-system layout optimization [49,50]

- Exhaust-system layout optimization, including outlet-noise assessment

- Valve-timing optimization

- Turbo-charging layout optimization including intercooler [51]

- Analysis and optimization of the engine transient behavior

- Extension to models for the simulation and optimization of the injection system, cam drivers, etc.

The reliability in the optimization process of the intake and exhaust system is appreciable in case of geometries with both a predominant one-dimensional flow and mixing processes of moderate complexity. Also preliminary investigations of the influence of valve-timing variations on the engine behavior can be performed successfully with a remarkable degree of freedom. Regarding the turbo-charging layout optimization, in particular under engine transient operation, there is none other simulation tool able to assist this relevant development task with such an efficiency and reliability. Especially this task, due to the resent increasing interest in turbo-charged engines, has boosted the implementation of the 1D-CFD-simulation in the engine development process.

The full-engine analysis simulation and the modular structure of this approach allows a convenient integration to other simulation programs (see Chapter 2.1) so that also other phenomena not directly related to the flow field analysis can be investigated (noise analysis, injection system optimization, etc.).

The main drawback of this approach is represented by the need of a complex model calibration supported by accurate experimental data, so that the base model can be successfully implemented for investigations based on parameter variations. This can be done as long as the implemented engine-process models have been developed and validated for the case analyzed.

6

Three-Dimensional Simulation (3D-CFD Simulation)

The three-dimensional simulation (3D-Computational-Fluid-Dynamics simulation) represents the most sophisticated approach for detailed numerical investigations on any thermo-fluid-dynamical problem. This approach is based on the numerical solution of partial differential equations for conservation of mass, species concentration, momentum and energy over an arbitrary fluid domain (Euler formulation). The basic formulations of the governing conservation equations are well known since the 19th century, but their numerical implementation in a computer-based simulation of a fluid-dynamical problem of interest is recent history.

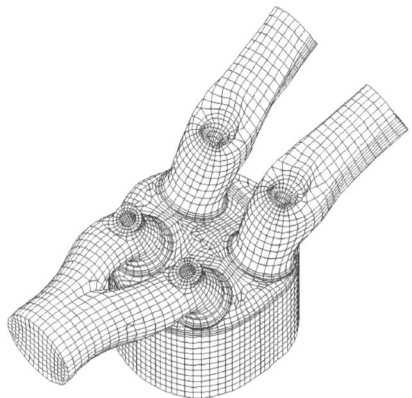

Figure 6.1: 3D-CFD-mesh of a cylinder of an internal combustion SI-engine.

With the advent of high-performance computers at the end of the eighties the numerical solution of complex fluid domains was made possible. To enable a computer for the solution of a continuum fluid domain (like the flow field inside the cylinder of an internal combustion engine), the continuum must be represented by a finite number of discrete elements (up to millions of elements). The most common method of discretization is to divide the fluid domain into a number of small zones or cells (finite volume approximation), which form a fine grid or mesh. The computational mesh (see Figure 6.1) serves as a framework for the local numerical solution of the discretized governing equations [52]. The time variable is similarly discretized into a sequence of small time intervals called time-steps, and, in case of unsteady flow, the transient solution is carried out in time: the solution at time t_{n+1} is calculated from the known solution at time t_n.

6.1   Fundamental Equations

The starting point for the 3D-CFD simulation are the conservation equations of fluid dynamics that represent the fundamental equations for any thermo-fluid-dynamical investigation. Since these are well established equations whose derivation is discussed in many books on fundaments of fluid dynamics [53-56], they are only briefly summarized here.

A conservation equation of an extensive variable $F(t)$ can be written in the following general form (Euler formulation):

$$\frac{\partial f}{\partial t} + div \, \vec{\Phi}_f = s_f + c_f \tag{6.1}$$

where $f(\vec{x},t) = dF / dV$ is the corresponding variable density or intensive variable of $F(t)$ in the volume element at the position $\vec{x}$. The conservation equation states that a change of the variable density $f(\vec{x},t)$ can be caused by a flux $\vec{\Phi}_f \cdot \vec{n} \, dS$ through the surface of the volume element, by a production or sink s_f and by long-range processes c_f. From this general formulation, conservation equations for mass, momentum, energy and species mass fractions can be derived. (More details regarding the vector and matrix analysis are found in Appendix A.)

6.1.1   Mass Conservation Equation

Considering the mass m of the volume element as the extensive variable $F(t)$ of interest, the density term $f(\vec{x},t)$ is here given by the mass density ρ and the flux density is the product of the local flow velocity $\vec{v}$ and ρ. Since mass can be neither formed nor destroyed due to engine processes, there are no production or long-range terms. From this it follows:

$$f = \rho \quad , \quad \vec{\Phi}_f = \rho \vec{v} \quad , \quad s_f = 0 \quad , \quad c_f = 0 \tag{6.2}$$

$$\frac{\partial \rho}{\partial t} + div\left(\rho \vec{v}\right) = 0. \tag{6.3}$$

6.1.2   Species Mass Conservation Equation

In case of the mass fraction $w_i = m_i/m$ of the species i, the density term is here given by the mass density ρ_i of species i. The local flow velocity $\vec{v}_i$ is composed of the mean flow velocity $\vec{v}$ and the diffusion velocity $\vec{V}_i$. The latter generates a species diffusion mass flux $\vec{j}_i$. Species can also be formed and destroyed in chemical reactions (e.g. combustion and dissociation processes), thus:

$$f = \rho_i = \rho\, w_i \; , \vec{\Phi}_f = \rho_i\, \vec{v}_i = \rho_i\left(\vec{v} + \vec{V}_i\right) = \rho_i\, \vec{v} + \vec{j}_i \; , s_f = M_i\, \omega_i \; , c_f = 0 \tag{6.4}$$

$$\frac{\partial \rho\, w_i}{\partial t} + \text{div}\left(\rho\, w_i \vec{v}\right) + div\, \vec{j}_i = M_i\, \omega_i \tag{6.5}$$

where M_i and ω_i are the molar mass and the molar formation rate of species i due to chemical reactions, respectively.

6.1.3   Momentum Conservation Equation
(Navier-Stokes' Equation)

In case of the conservation of momentum, the density to be substituted in the general form is given by the momentum density $\rho \vec{v}$, the momentum flux consists of a convective part $\rho \vec{v} \otimes \vec{v}$ and the second-order stress tensor $\overline{\overline{P}}$ (matrix) which describes the momentum change due to viscous effects $\overline{\overline{\Pi}}$ (shear-stress tensor) and the pressure p; also a long-range term $\rho \vec{g}$ (gravitation) can be taken into account, thus:

$$f = \rho \vec{v} \quad , \quad \vec{\Phi}_f = \rho\, \vec{v} \otimes \vec{v} + \overline{\overline{P}} \quad , \quad s_f = 0 \quad , \quad c_f = \rho \vec{g} \tag{6.6}$$

where:

$$\overline{\overline{P}} = p\overline{\overline{I}} + \overline{\overline{\Pi}} \tag{6.7}$$

so that:

$$\frac{\partial(\rho\,\vec{v})}{\partial t} + \text{div}\left(\rho\,\vec{v}\otimes\vec{v}\right) + div\,\overline{\overline{\Pi}} - grad\,p = \rho\vec{g}\;.\tag{6.8}$$

6.1.4 Energy Conservation Equation

The conservation equation of energy can be derived in various forms. In the most general formulation the energy amount is composed of the internal energy u, the kinetic energy, the potential gravitational energy G, and the heat of formation of the mixture h_f (chemical energy) respectively:

$$\rho e = \rho\left(u + \frac{1}{2}|\vec{v}|^2 + G + h_f\right).\tag{6.9}$$

Since energy can be neither formed nor destroyed due to engine processes, there is no production term s_f. The long-range term is used to take into account, e.g., the effects q_r of radiation or magnetic fields. From this it follows:

$$f = \rho e \quad , \quad \vec{\Phi}_f = \rho e\vec{v} + \overline{\overline{P}}\vec{v} + \vec{j}_q \quad , \quad s_f = 0 \quad , \quad c_f = q_r \tag{6.10}$$

where the energy flux $\vec{\Phi}_f$ is composed of the convective term $\rho e\vec{v}$, a term $\overline{\overline{P}}\vec{v}$ which represents the energy transport due to pressure and shear stresses (viscosity diffusion) and the energy transport $\vec{j}_q$ in particular due to heat conduction (see Chapter 6.2.1.2).

In the 3D-CFD simulation of internal combustion engines it is customary to adopt the static thermal-chemical enthalpy h_{tc}, that embodies the thermal h and the chemical contribution h_f, respectively, as the variable characterizing the energy content of the fluid. It follows:

$$h_{tc} = h + h_f \tag{6.11}$$

$$\frac{\partial(\rho h_{tc})}{\partial t} - \frac{\partial p}{\partial t} + \text{div}\left(\rho h_{tc}\vec{v} + \vec{j}_q\right) + \overline{\overline{P}} : \text{grad}(\vec{v}) - div(p\vec{v}) = q_r\;.\tag{6.12}$$

6.2 Engine Modeling

The fundamental equations of the fluid dynamics and the thermal state equation of the working fluid described in Chapter 6.1, build an equation system of $6 + N_l$ scalar equations, for each finite volume, where N_i is the total number of described species in the gas mixture. In order to close this system, several laws or models, mostly based on empirical formulations, used to describe e.g. flux densities ($\vec{j}_i$, $\vec{j}_q$, etc.), the pressure tensor $\overline{\overline{P}}$, thermo-physical properties of the fluid (ρ, h, etc.), mainly as a function of the physical properties (ρ, T, $\vec{v}$ and w_i) are required. The above mentioned laws or models perform the task of converting the physical problem into a mathematical formulation. This is a critical step in the simulation that requires, especially for an extremely complex phenomenon, to accurately identify the interactions between *causes* – e.g. described by the physical properties of the fluid - and *effects* – e.g. flux density – so that a mathematical formulation can be developed.

These mathematical 3D-models can be divided into two categories: *universally-valid models* and *engine-specific models* depending on whether the models have general validity or their implementations are limited to the simulation of internal combustion engines only.

6.2.1   Universally-Valid 3D-CFD-Models

The 3D-CFD-models presented in this paragraph are supposed to have general validity in any fluid domain. As introduced before, these models are mostly based on empirical formulations. The mathematical formulation of each single physical phenomenon has usually a good agreement with "fundamental" experimental investigations, with some exceptions regarding spray atomization, diffusive combustion, pollution formation and mainly all unsteady thermodynamic processes within the boundary layers (near-wall region). Focusing on the mathematical formulation of a complex fluid-dynamical problem of interest, like that of an internal combustion engine in which countless physical processes take place, the general accuracy of the implemented 3D-CFD-models, apart from the mesh influence, depends principally on both the assumptions and simplifications made to describe each relevant single physical phenomenon.

Due to simplicity this paragraph will focus on the following models: description of the thermo-physical properties state of the working fluid for a complex mixture, non-convective processes, turbulence, combustion and wall heat transfer.

6.2.1.1   Modeling of the Thermo-physical Properties of the Working Fluid

Like in the real working-process analysis, the well known thermal state equation of general formulation – *the perfect gas equation of state* - is used to relate temperature T, pressure p and density $\rho = m/V$ of the working fluid in the cylinder:

$$pV = mRT. \tag{6.13}$$

In the definition of the real gas constant R in case of a gas mixture this term is described as follows:

$$R = \frac{\Re}{M} = \Re \cdot \sum_i \frac{w_i}{M_i}. \tag{6.14}$$

When the temperature in the fluid T is nearly equal or lower than the critical temperature T_{cr} or when the pressure p is nearly equal or higher than the critical pressure p_{cr}, the density of the fluid is inadequately predicted by using the perfect gas equation of state (see Eq. 6.13). Since the pressure and temperature ranges in the simulation of internal combustion engines are far away from the critical conditions T_{cr} and p_{cr} the perfect gas equation of state ensures a high accuracy in predicting the fluid density and it is not required to take real gas effects into account. A relationship for ideal gases is then used to link the specific internal energy u with the specific enthalpy h (thermal contribution):

$$h = u + RT. \tag{6.15}$$

The dependence of the thermal enthalpy h on the temperature is non-linear. For numerical applications it is convenient to model h using the following polynomial functions:

$$h = \sum_i w_i \cdot h_i = RT \left(\sum_{j=1}^{5} \frac{a_j}{j} T^{j-1} + \frac{a_6}{T} \right) \tag{6.16}$$

where:

$$a_j = \sum_i w_i \cdot a_{j,i}. \tag{6.17}$$

The coefficients $a_{j,i}$ for each species i are usually tabulated into detailed tables (see e.g. JANAF-Tables [57] or CHEMKIN-Tables). In a similar way the heat of formation of the mixture h_f (chemical contribution) is described as follows:

$$h_f = \sum_i w_i \cdot h_{f,i} \tag{6.18}$$

where $h_{f,i}$ is the standard enthalpy of formation of each species i.

6.2.1.2   Modeling of Non-Convective Processes

The following models deal with all energy exchange processes (non-convective processes) that are not directly related to energy transport by fluid motion (convective processes) [55,58].

Fick's Law

The diffusion mass flow $\vec{j}_i$ of the species i is assumed to be caused by three effects:

$$\vec{j}_i = \vec{j}_i^d + \vec{j}_i^T + \vec{j}_i^p \qquad (6.19)$$

where the three contributions are: *ordinary diffusion* $\vec{j}_i^d$ due to species concentration gradient, *thermal diffusion* $\vec{j}_i^T$ (Soret-effect) due to temperature gradient and *pressure diffusion* $\vec{j}_i^p$ which is caused by pressure gradient. Fick found out empirically that the following mathematical formulations for $\vec{j}_i^d$ and $\vec{j}_i^T$ are adequate to describe the physical processes of ordinary and thermal diffusion, respectively:

$$\vec{j}_i^d = -D_i^M \rho \frac{w_i}{x_i} grad \, x_i \approx -D_i^M \rho \; grad \, w_i \qquad (6.20)$$

$$\vec{j}_i^T = -D_i^T \cdot grad \left(\ln T \right) \qquad (6.21)$$

where $x_i = n_i/n$ is the mole fraction of the species i. D_i^M denotes the diffusion coefficient for the diffusion of the species i into the mixture, which represents the sensitivity of the gas mixture to generate a diffusion mass flow in consequence of a concentration gradient. The coefficient D_i^T is the analog of D_i^M for thermal diffusion. The pressure diffusion $\vec{j}_i^p$ in the characteristic range of the pressure of internal combustion engines can be neglected.

Fourier's Law

The heat flux $\vec{j}_q$ is also assumed to be caused by three effects:

$$\vec{j}_q = \vec{j}_q^c + \vec{j}_q^d + \vec{j}_q^D \qquad (6.22)$$

where the three contributions here are the *heat conduction* $\vec{j}_q^c$ due to a temperature gradient, the *diffusion heat* $\vec{j}_q^d$ due to a diffusion mass flux $\vec{j}_i$ of each species i and the *Dufour-effect* $\vec{j}_q^D$ which represents the heat transfer caused directly by the species concentration gradients. Fourier found empirically that the following mathematical formulation for $\vec{j}_q^c$ is adequate to describe the physical processes of the heat conduction:

$$\vec{j}_q^c = -\lambda \cdot grad \, T \qquad (6.23)$$

where λ is the thermal conductivity, usually approximated by a polynomial expression as a function of the temperature $\lambda = f(T)$. The thermal conductivity λ represents the sensitivity of both the gas mixture and a solid to generate a molecular heat flux in consequence of a temperature gradient. The diffusion heat $\vec{j}_q^d$ due to the mass diffusion becomes:

$$\vec{j}_q^d = \sum_i h_{tc,i} \, \vec{j}_i .$$

(6.24)

The heat flux $\vec{j}_q^D$ due to the Dufour-effect can also be neglected in the simulation of internal combustion engines.

Newton's Law

Many experimental investigations on viscous fluids directed by Newton at the end of the 17th century led for the shear-stress tensor $\overline{\overline{\Pi}}$ to the following empirical formulation:

$$\overline{\overline{\Pi}} = -\mu \left[(grad \, \vec{v}) + (grad \, \vec{v})^T - \frac{2}{3} \cdot (div \, \vec{v}) \cdot \overline{\overline{I}} \right] .$$

(6.25)

The dependence of the temperature on the dynamic viscosity μ is usually obtained by the experimental formulation:

$$\frac{\mu}{\mu_0} = \left(\frac{T}{T_0} \right)^{0.7}$$

(6.26)

where μ_0 is the dynamic viscosity at standard temperature $T_0 = 273$ K.

6.2.1.3   Turbulence Modeling

In laminar flows, vectorial and scalar quantities have well-defined values. In contrast turbulent flows in a space defined by a characteristic length l, are characterized by continuous chaotic fluctuations of velocity leading to fluctuations in scalars such as density ρ, temperature T, mixture composition w_i, etc. (see Figure 6.2).

The fluctuations of these turbulent flows, which are of interest in internal combustion engines, are caused by vortices that are generated by shear stresses within the flow. The growth of these vortices is the result of a "competition" between nonlinear generation processes due to the kinetic energy of the fluid and dumping processes caused by viscous dissipation ($\sim$ to the dynamic viscosity μ). The influence of the generation processes overtakes that one of the dumping processes when a critical value of the Reynolds number Re is exceeded (see Eq. 6.27). This value ($Re \cong 2.300$ for a flow in a pipe with diameter l) represents the transition condition from laminar to turbulent behavior of the flow.

$$Re = \frac{\rho |\bar{v}| l}{\mu} \tag{6.27}$$

The character of a turbulent flow depends on its environment. In the cylinder, the flow is unsteady and shows substantial cycle-by-cycle fluctuations. An important characteristic of a turbulent flow is its irregularity or randomness. Statistical methods are therefore helpful to describe such a flow field. One approach used in quasi-periodic flows is the so called *ensemble-averaging* or *phase-averaging*, where e.g. the ensemble averaged velocity component $\bar{U}(\varphi)$ of the velocity vector $\bar{v}$ (see Figure 6.2) is the average of the measured velocity components over a large number of operating cycles N_C at the same crank angle φ :

$$\bar{U}(\varphi) = \frac{1}{N_C} \sum_{i=1}^{N_C} u(\varphi, i). \tag{6.28}$$

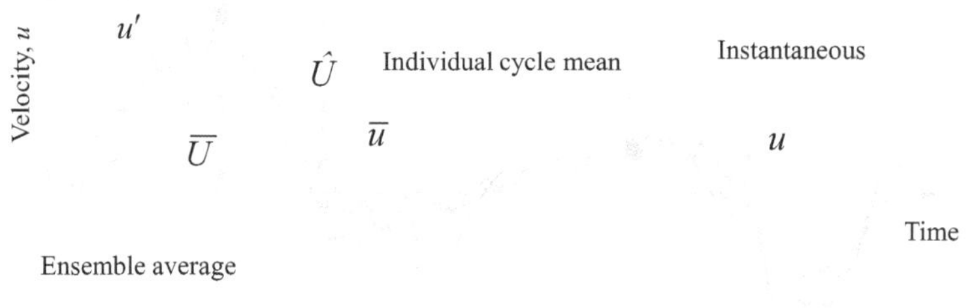

Figure 6.2: *Velocity profile at a fixed position in the cylinder during an engine cycle.*

A mean value of the velocity component $\bar{u}(\varphi, i)$ during each individual cycle i can be defined as follows:

$$\bar{u}(\varphi, i) = \frac{1}{\Delta\varphi} \int_{\varphi}^{\varphi+\Delta\varphi} u(\varphi, i) \cdot d\varphi \tag{6.29}$$

where the integration period $\Delta\varphi$ must be greater than the characteristic time of the turbulent fluctuations but also sufficiently smaller to detect the temporal variation of the mean value. The turbulent intensity component $u'(\varphi, i)$ of the velocity fluctuations is then defined as:

$$u'(\varphi, i) = \frac{1}{\Delta\varphi} \left\{ \int_{\varphi}^{\varphi+\Delta\varphi} \left[u^2(\varphi, i) - \bar{u}^2(\varphi, i) \right] \cdot d\varphi \right\}^{1/2}. \tag{6.30}$$

More than the turbulent intensity component $u'(\varphi,i)$ it is customary to introduce the turbulent kinetic energy (TKE) k as the variable characterizing the turbulence degree of the flow. Assuming an isotropic turbulence in the flow (the three components of the turbulent velocity are approximately equal), it follows:

$$k = \frac{3}{2}u'^2 .$$
(6.31)

The difference between the mean velocity in a particular cycle $\bar{u}(\varphi,i)$ and the ensemble-averaged mean velocity $\bar{U}(\varphi)$ is defined as the cycle-by-cycle variation in the mean velocity:

$$\hat{U}(\varphi,i) = \bar{U}(\varphi) - \bar{u}(\varphi,i).$$
(6.32)

In turbulent flows, some length scales are helpful in order to characterize different aspects of the flow behavior (see Figure 6.3).

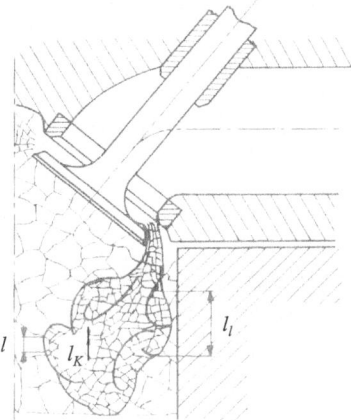

Figure 6.3: *Schematic of the turbulent structure during the intake phase.*

The dimension of the largest turbulent eddies in the flow (lower perturbation frequency) – integral length scale l_l (~ 3-4 mm for IC-engines) – are limited in size by the geometrical boundaries of the system. In contrast, the dimension of the smallest turbulent eddies (higher perturbation frequency) – Kolmogorov length scale l_k (~ 0.015 mm for IC-engines as an average over the operating cycle) – are limited by the scale of molecular diffusion at which the turbulent kinetic energy k dissipates directly into internal energy by viscosity (see Eqs. 6.52 and 6.53).

The definition of the Kolmogorov length l_k is extremely important for CFD-simulations (see Chapter 7.1.1.1). It represents the dimension of the smallest eddies of the turbulent field. If the domain is discretized using cells with lengths much smaller than l_k (Direct Numerical Simulation-approach) a numerical and detailed solution of the flow field without any turbulence modeling can be performed. As it will be explained in Chapter 7.1.1.1 this kind of simulation is not of practical interest (e.g. the simulation of an internal combustion engine) because the required computational resources would probably exceed the capacity of the most powerful computers currently available (status 2009).

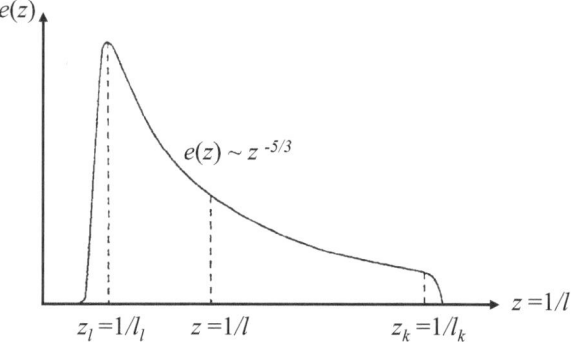

Figure 6.4: *Spectrum of turbulent kinetic energy as a function of turbulent scale.*

The distribution of the turbulent kinetic energy $k(\bar{x},t)$ at one position in the fluid domain among the spectrum of turbulent eddies with diameter l ($z = 1/l$) is described by the *turbulent energy spectrum* (see Figure 6.4) This spectrum shows an *energy cascade* of the turbulent energy density $e(y,\bar{x},t)$ from large eddies l_l, in which is stored the major part of the kinetic turbulent energy, to many small eddies.

$$k(\bar{x},t) = \int_0^\infty e(z,\bar{x},t)\,dz \ . \tag{6.33}$$

Favre-Averaging

With exception of the DNS-approach, in order to solve the conservation equations for turbulent flows, especially when combustion processes with large density variations take place, it is useful to introduce another average, called *Favre-average*. The Favre-average, for an arbitrary property q, is given by:

$$\tilde{q} = \frac{\overline{\rho q}}{\overline{\rho}} \tag{6.34}$$

so that:

$$q(\vec{x},t) = \overline{q}(\vec{x},t) + q'(\vec{x},t) = \tilde{q}(\vec{x},t) + q''(\vec{x},t). \tag{6.35}$$

The main reason for the use of the Favre-average (also called density-weighted average) is a much more compact mathematical formulation of the fundamental equations of the fluid dynamics when, in case of turbulent flows, with exception of the DNS-simulation approach (see Chapter 7.1.1.1), it is necessary to split the variables into their mean and fluctuation values. E.g. the averaging of the term ρuv by the Favre-averaging requires the following more compact formulation than by the weighted averaging:

$$\overline{\rho uv} = \overline{\rho} \cdot \tilde{u} \cdot \tilde{v} + \overline{\rho u''v''} = \overline{\rho} \cdot \overline{u} \cdot \overline{v} + \overline{\rho} \cdot \overline{u'v'} + \overline{u} \cdot \overline{\rho'v'} + \overline{v} \cdot \overline{\rho'u'} + \overline{\rho'u'v'}. \tag{6.36}$$

The fundamentals equations for turbulent flows after Favre-averaging become:

$$\frac{\partial \overline{\rho}}{\partial t} + \mathrm{div}\left(\overline{\rho}\tilde{v}\right) = 0 \tag{6.37}$$

$$\frac{\partial \left(\overline{\rho}\tilde{w}_i\right)}{\partial t} + \mathrm{div}\left(\overline{\rho}\,\tilde{w}_i\tilde{v}\right) + \mathrm{div}\left(\overline{\rho\,w_i''\,\overline{v}''}\right) + \mathrm{div}\,\overline{\overline{j}}_i = \overline{M_i\omega_i} \tag{6.38}$$

$$\frac{\partial \left(\overline{\rho}\tilde{v}\right)}{\partial t} + \mathrm{div}\left(\overline{\rho}\,\tilde{v}\otimes\tilde{v}\right) + \mathrm{div}\left(\overline{\rho\,\overline{v}''\otimes\overline{v}''}\right) + \mathrm{div}\overline{\overline{\Pi}} - \mathrm{grad}\,\overline{p} = \overline{\rho}\tilde{g} \tag{6.39}$$

$$\frac{\partial \left(\overline{\rho}\tilde{h}\right)}{\partial t} - \frac{\partial \overline{p}}{\partial t} + \mathrm{div}\left(\overline{\rho}\tilde{h}\tilde{v} + \overline{\overline{j}}_q\right) + \mathrm{div}\left(\overline{\rho h''\overline{v}''}\right) + \overline{\overline{P}:\mathrm{grad}(\overline{v})} - \mathrm{div}\left(\overline{p\overline{v}}\right) = \tilde{q}_r \tag{6.40}$$

where new terms depending on the fluid turbulent fluctuations appear. These terms introduce new unknown variables so that the equation system becomes *unclosed*. Actually, turbulence modeling means, first of all, development of mathematical formulations, that link the turbulence terms $\overline{j}_{i,T}^d$, $\overline{\overline{\Pi}}_T$ and $\overline{j}_{q,T}^d$ to the averaged property of the fluid in an analogous way to the modeling of the molecular processes $\overline{j}_i^d$, $\overline{\overline{\Pi}}$ and $\overline{j}_q^c$, thus:

$$\overline{j}_{i,T}^d = \overline{\rho\,w_i''\,\overline{v}''} = -D_{i,T}^M\,\overline{\rho}\,\mathrm{grad}\,\tilde{w}_i \tag{6.41}$$

$$\overline{\overline{\Pi}}_T = \overline{\rho\,\overline{v}''\otimes\overline{v}''} = -\mu_T\cdot\left[\left(\mathrm{grad}\,\tilde{v}\right) + \left(\mathrm{grad}\,\tilde{v}\right)^T\right] \tag{6.42}$$

$$\overline{j}_{q,T}^c = \overline{\rho\,h''\,\overline{v}''} - -\lambda_T\,\mathrm{grad}\,\tilde{T} \tag{6.43}$$

where $D_{i,T}^{M}$, μ_T and λ_T are called *turbulent exchange coefficients*. In an internal combustion engine due to its geometry and the rapidity at which the different phases succeed, the flow motion is strongly turbulent. Consequently the processes occurring are predominately driven by the turbulent exchange terms and the molecular exchange processes can be very often neglected.

Now a formulation for each turbulent exchange coefficient is required. This task is the "core" of the turbulence modeling.

$\tilde{k} - \tilde{\varepsilon}$ - Turbulence Modeling

The most known turbulence model is represented by the $\tilde{k} - \tilde{\varepsilon}$ -approach (for more information about other turbulence models see [5,37,55,56,58,59]). In this model the turbulent exchange coefficients are expressed in terms of the turbulent kinetic energy $\tilde{k}$ and the dissipation rate $\tilde{\varepsilon}$ of the turbulent kinetic energy:

$$\tilde{\varepsilon} = \frac{d\tilde{k}}{dt} \tag{6.44}$$

$$D_{i,T}^{M} = \frac{\mu_T}{Sc\,\overline{\rho}} \tag{6.45}$$

$$\mu_T = f_\mu \frac{C_\mu\,\overline{\rho}\,\tilde{k}^2}{\tilde{\varepsilon}} \tag{6.46}$$

$$\lambda_T = \frac{\mu_T\,c_p}{Pr} \tag{6.47}$$

where:

$$Pr = \frac{c_p\,\mu}{\lambda} \quad \text{and} \quad Sc = \frac{\mu}{\overline{\rho}\,D_i^M} \tag{6.48}$$

are the Prandtl and Schmidt numbers, respectively.

The Prandtl number defines the ratio of momentum diffusivity due to viscosity effects and thermal diffusivity and the Schmidt number defines the ratio of momentum diffusivity and mass diffusivity. The following coefficients f_μ and C_μ (see Table 6.1) refer to the fluid region where the turbulence is fully developed (high-Reynolds number domain). This region belongs to a fluid domain outside the near-wall region where the viscous effects become negligible so that the turbulence can be assumed isotropic (More details regarding the turbulence in the near-wall region in Chapter 10). In order to close the equation system two empirical transport equations for $\tilde{k}$ and $\tilde{\varepsilon}$ are defined as follows:

$$\frac{\partial(\overline{\rho}\widetilde{k})}{\partial t} + \mathrm{div}\left(\overline{\rho}\,\widetilde{k}\,\widetilde{v}\right) - \frac{\mu + \mu_T}{\sigma_k}\cdot\nabla^2\widetilde{k} =$$

$$= \mu_T\left\{\left[\left(grad\,\widetilde{v}\right)+\left(grad\,\widetilde{v}\right)^T\right]:\left[\left(grad\,\widetilde{v}\right)-\left(grad\,\widetilde{v}\right)^T\right]\frac{grad(\overline{\rho})}{Pr\cdot\overline{\rho}}\right\} +$$

$$-\frac{2}{3}\left(\overline{\rho}\widetilde{k} + \mu_T\ div\,\widetilde{v}\right)\cdot div\,\widetilde{v} - \overline{\rho}\,\widetilde{\varepsilon} \tag{6.49}$$

$$\frac{\partial(\overline{\rho}\widetilde{\varepsilon})}{\partial t} + \mathrm{div}\left(\overline{\rho}\,\widetilde{\varepsilon}\,\widetilde{v}\right) - \frac{\mu + \mu_T}{\sigma_\varepsilon}\cdot\nabla^2\widetilde{\varepsilon} = A \tag{6.50}$$

where σ_k and σ_ε are also empirical coefficients. In the $\widetilde{k} - \widetilde{\varepsilon}$ - standard turbulence model, the term A in the transport equation of the dissipation rate $\widetilde{\varepsilon}$ of the turbulent kinetic energy becomes:

$$A = C_{\varepsilon 1}\cdot\frac{\widetilde{\varepsilon}}{\widetilde{k}}\left\langle\mu_T\left\{\left[\left(grad\,\widetilde{v}\right)+\left(grad\,\widetilde{v}\right)^T\right]:\left[\left(grad\,\widetilde{v}\right)-\left(grad\,\widetilde{v}\right)^T\right]+\right.\right.$$

$$\left.\left.- C_{\varepsilon 3}\frac{grad(\overline{\rho})}{Pr\,\overline{\rho}}\right\} - \frac{2}{3}\left(\overline{\rho}\widetilde{k} + \mu_T\ div\,\widetilde{v}\right)\cdot div\,\widetilde{v}\right\rangle - C_{\varepsilon 2}\overline{\rho}\frac{\widetilde{\varepsilon}^2}{\widetilde{k}} - C_{\varepsilon 4}\overline{\rho}\,\widetilde{\varepsilon}\ div\,\widetilde{v}\ . \tag{6.51}$$

The empirical coefficients of this turbulence model are reported below (see Table 6.1):

Table 6.1: Empirical coefficients of the $\widetilde{k} - \widetilde{\varepsilon}$ standard turbulence model (Fully developed turbulence).

f_μ	C_μ	σ_k	σ_ε	$C_{\varepsilon 1}$	$C_{\varepsilon 2}$	$C_{\varepsilon 3}$	$C_{\varepsilon 4}$
0.95	0.09	1.0	1.22	0.44	0.92	0.44	0.33

From the previous relations the integral length scale l_l and the Kolmogorov length scale l_k can be defined as follows:

$$l_l = C_\mu^{\frac{3}{4}}\cdot\frac{\widetilde{k}^{\frac{3}{2}}}{\widetilde{\varepsilon}} \tag{6.52}$$

$$l_k = \left(\frac{(\mu/\rho)^3}{\widetilde{\varepsilon}}\right)^{\frac{1}{4}} \tag{6.53}$$

Reliability of Turbulence Models in the Simulation of IC-Engines

In order to describe the turbulent exchange coefficients, several turbulence models based on different approaches and different mathematical complexity have been developed. Due to the "evident" complexity of turbulent processes, all existing turbulence models are rough representations of the physical phenomena involved. This is a generally recognized fact, which nonetheless deserves mention here. It is also known that the degree of approximation in a given turbulence model depends on the nature of the flow to which both it is being applied or has been validated, and that the characterization of the circumstances which give rise to 'good' or 'inaccurate' performance must unfortunately be based mainly on experience.

In case of internal combustion engines turbulent phenomena, due to high temperature, velocity and compositions gradients, reach complex transient structures that are barely comparable to that of other thermodynamic machines. Actually, there are no devices able to reliably measure the turbulence within the combustion chamber, i.e. a validation procedure of turbulence models under real engine conditions is not possible. This introduces a very sensible factor in the reliability of simulation results, because, as well known, turbulence is the most important "driving variable" of engine processes (combustion, wall heat transfer, mixture formation, etc.). Due to previous considerations it would be confusing and probably misleading to try analyzing and improving turbulence models on considerations of a local turbulent field; in particular it is well known that the mesh structure, more than the discretization degree, drastically influences the calculated turbulence. Therefore, at the end it makes more sense to evaluate the turbulence models from a global point of view, i.e., whether the global results of the 3D-CFD-models "feeded" by turbulence and directly responsible for the simulation of engine operation (see Chapter 2.3) are in accordance with both experimental measurements and other simulation programs (e.g. the working-process analysis) or not. Of course here a separation of the contributions to the results reliability between the involved process and the turbulence as input variable is extremely difficult, but in fact no other more promising approaches are available.

6.2.1.4   Combustion Models

Summarily, the simulation of reacting flows requires one more ingredient in order to provide the connection between the fluid's mechanical and chemical behavior: in Eq. (6.5) information are needed for the chemical source term $r_i = M_i\omega_i$ of each species involved in the chemical reactions. Reaction mechanisms in internal combustion engines, which involve hydrocarbon combustion, are a highly complex system with thousands of species and hundreds of competing reactions. Moreover, the exact chemistry of the combustion processes and the way in which one reaction step influences another is not well understood yet. Thus, simple "artificial" reaction schemes are often set as approximations to the more complex ones, to make them easier to be analyzed. Several of this simplified reaction schemes has been proposed, but still exorbitant CPU-time has limited their application to fundamental research investigations (more details in Chapters 7, 8 and 9).

The previous considerations explain the reason why in case of a 3D-CFD-simulation of practical interest, it is more opportune and convenient to implement engine specific combustion models that actually are limited to heat-release prediction, skipping a detailed analysis of the combustion mechanisms which are not relevant in the prediction of the engine operating cycle (see Chapter 6.2.2).

6.2.1.5 Wall Heat-Transfer Models

The temperature difference between the fluid and the wall generates a heat-transfer process that has a remarkably influence on the engine operating cycle (see Chapter 3). In this near-wall region the flow field is, especially in case of internal combustion engines, extremely complex. Approaching the wall the temperature changes drastically from the fluid in the region of fully turbulence development (isotropic turbulence – see Chapter 6.2.1.3) through a turbulent anisotropic flow field with increasing laminarization degree and a fully laminar flow in the wall-near region up to the wall itself, where there is no molecular motion due to friction. This flow field is described by the boundary layer theory that, in case of a simple geometry and stationary flow, delivers equations for the prediction of the temperature and the velocity profile (coupling between the thermal boundary layer and the momentum boundary layer) within the near-wall region. Starting from the temperature gradients, it is possible then to calculate the corresponding wall heat-transfer.

In the 3D-CFD-simulation the calculation of the wall heat-transfer is commonly based on the standard "wall function approach". This approach estimates the temperature and velocity profiles of the boundary layer according to the theory of a streamed horizontal plate in case of a stationary flow. Since this assumption is far away from the real transient boundary layer that is locally found in a combustion chamber, the reliability of the results is very low (more details in Chapter 10).

6.2.2 Introduction to Engine-Specific 3D-CFD-Models

In theory there is no need to implement models in 3D-CFD-codes which are dedicated to the reproduction of phenomena that take place in internal combustion engines. A solution of the equation system should also be possible using the fundamental equations in combinations with models of general validity that have been introduced before. Since the phenomena are extremely complex and often not adequately understood at a fundamental level, i.e. it is not possible to find a mathematical formulation based on both basic governing equations and empirical laws of general validity, it is necessary to introduce 3D-models that are supposed to be valid only for the simulation of internal combustion engines. This is an irreplaceable process that allows, using phenomenological approaches, empirical relations and "ad hoc" approximations, to bridge the gaps in the understanding of critical phenomena [10,11,23,36,60].

In comparison to the models of general validity, engine-specific models are not only functions of the local physical properties of the fluid but also of the engine relevant parameters. Moreover, in some cases, it is also convenient to substitute universally-valid models with engine-specific models of a more simplified and "robust" formulation so that a reduction of the CPU-time and an increase of the accuracy of the results can be performed.

The following short list reports the main engine processes that can be described by engine-specific 3D-CFD-models. A few of them will be discussed in details in Chapters 8, 9 and 10:

- modeling of the thermo-physical properties of the working fluid

- spark ignition modeling

- combustion modeling for spark ignition and diesel engines

- self-ignition modeling

- wall heat transfer modeling

- injector spray modeling

6.3   Discretization Practices (Numerical Implementation)

The system of non-linear partial equations – that represent the mathematical description of the physical problem of the fluid dynamics introduced in this chapter – cannot be analytically solved. In order to solve this system, some methods (discretization practices) which allow to approximate the solution in the continuum using a discretized flow field (computational mesh - see Figure 6.1) are required. Actually, the discretization practice transforms the non-linear differential equations of the mathematical model into algebraic equations and employs solution algorithms to obtain the dependent parameters from the discretized equations (numerical solution) [53,56]. In the following, a brief description of the discretization practices is reported.

In the finite volume method the flow parameters are approximated in terms of the cell-centered nodal values, i.e. considering the discrete element in Figure 6.5 it is assumed that the dependent parameters do not vary within the cell control-volume $V_{\widetilde{c}}$ and that they are concentrated in the central node P. The conservation equations are first integrated over the varying cell control-volume $V_{\widetilde{c}}(t)$ and the divergence terms are transformed into surface integrals by using the Gauss divergence theorem. Considering the conservation equation of an arbitrary variable $f(\vec{x},t)$ (see Eq. 6.1), it is possible to write its correspondent integral form:

$$\frac{\partial}{\partial t} \int_{V_{\widetilde{c}}} f \cdot dV + \int_{\partial V_{\widetilde{c}}} \vec{\Phi}_f \cdot \vec{n} \cdot dA = \int_{V_{\widetilde{c}}} (s_f + c_f) \cdot dV \qquad (6.54)$$

where $\partial V_{\widetilde{c}}$ is the closed surface which envelopes the cell.

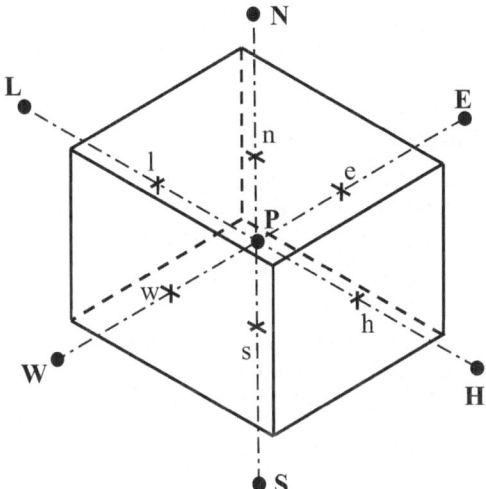

Figure 6.5: *Discretization practice over a discrete element (cell).*

From here onwards some approximations have to be introduced in order to transform the integral relations into algebraic equations. The first term can be approximated assuming a homogeneous function $f(\vec{x},t)$ over the volume $V_{\widetilde{c}}(t)$, so that the integral function is no longer required. Successively the derivative can be replaced with its incremental ratio, thus:

$$\frac{\partial}{\partial t} \int_{V_{\widetilde{c}}} f \cdot dV \cong \frac{f(\vec{x}_P, t + \Delta t) \cdot V_C(t + \Delta t) - f(\vec{x}_P, t) \cdot V_{\widetilde{c}}(t)}{\Delta t}. \qquad (6.55)$$

The second term, which represents the flux of the variable $f(\bar{x},t)$ through the surface $\partial V_{\bar{C}}$ of the cell, can be expressed in terms of average values of the variable $f(\bar{x},t)$ and the velocity $v(\bar{x},t)$ over each surface A_j with the normal vector $\bar{n}(\bar{x}_j,t)$ of the cell, thus:

$$\int_{\partial V_{\bar{C}}} \bar{\Phi}_f \cdot \bar{n} \cdot dA = \sum_j \left(f(\bar{x}_j,t) \cdot \bar{v}(\bar{x}_j,t) \cdot \bar{n}(\bar{x}_j,t) \right) \cdot A_j \qquad j = N, W, S, E, L, H \tag{6.56}$$

where:

$$\partial V_{\bar{C}} = \sum_j A_j \qquad j = N, W, S, E, L, H \ . \tag{6.57}$$

The third term, analog to the first one, becomes:

$$\int_{V_{\bar{C}}} (s_f + c_f) \cdot dV = \left(s_f(\bar{x}_P,t) + c_f(\bar{x}_P,t) \right) \cdot V_{\bar{C}} \ . \tag{6.58}$$

6.3.1   Spatial Flux Discretization

As it has been seen in Eq. 6.56, the discretization of the flux terms in the conservation equations requires expressing the values of the variables $f(\bar{x}_j,t)$ at the surface positions of the cells. The manner in which the fluxes of the generic variable $f(\bar{x},t)$ are expressed in terms of nodal values $f(\bar{x}_P,t)$ is one of the key factors determining the accuracy and the stability in the 3D-CFD simulation. There are essentially two main classes of flux approximation in widespread use, namely *low-order* and *higher-order* differencing schemes.

6.3.1.1   Low-Order Differencing Scheme – Upwind Differencing (UD)

The value of the variables $f(\bar{x}_j,t)$ at the surface position is selected from the nearest upwind neighbor central-node value, e.g. in case of surface H (see Figure 6.5):

$$f(\bar{x}_H,t) = \begin{cases} f(\bar{x}_P,t) & if \quad \bar{v}(\bar{x}_H,t) \cdot \bar{n}(\bar{x}_H,t) \geq 0 \\ f(\bar{x}_H,t) & if \quad \bar{v}(\bar{x}_H,t) \cdot \bar{n}(\bar{x}_H,t) < 0 \end{cases} \tag{6.59}$$

This differencing scheme, which characteristically generates discretized equation forms that are easy to solve and CPU-time convenient, produces solutions which obey the expected physical bounds, but sometimes gives rise to smearing of gradients. The latter effect has come to be known as *numerical diffusion*. This is a form of a truncation error that diminishes as the grid is refined, but with a disproportionate increase of CPU-time (see Chapter 7.1.1).

6.3.1.2   Higher-Order Differencing Scheme

Several higher-order differencing schemes have been proposed and implemented in commercial 3D-CFD-codes. By means of an example – the central differencing scheme (CD) – the basic concept of these differencing schemes will be introduced. The CD-scheme belongs to the category of second-order differencing schemes, which simply interpolate linearly on the nearest neighbor values, irrespective of the flow direction, giving in case of surface *H* (see Figure 6.5):

$$f(\vec{x}_h, t) = g \cdot f(\vec{x}_P, t) + (1 - g) \cdot f(\vec{x}_H, t) \tag{6.60}$$

where g is a geometric interpolation factor. Other higher-order differencing schemes are, e.g. linear upwind differencing (LUD), self-filtered central differencing (SFCD), Gamma differencing, blended differencing (BD), quadratic upstream interpolation of convective kinematics (QUICK) and monotone advection and reconstruction scheme (MARS) [53,54,56, 61].

Higher-order differencing schemes, usually better preserve steep gradients, but often they lead to equations that are more difficult to solve (and, in extreme cases, they may provoke numerical instabilities). These schemes, also may lead to values without physical meaning (e.g. negative turbulent kinetic energy, negative species' mass fraction, etc.). This phenomenon is often called *numerical dispersion*. This phenomenon can be partly diminished by grid refinement or by implementing blending factors in the differencing schemes.

6.4   The Role of the 3D-CFD-Simulation in the Engine Development Process

The 3D-CFD-simulation represents indisputably the most detailed approach for the investigation of the engine operating cycle. From its basic concept this approach should allow an unlimited predictability of engine processes by unrestrictedly varying of any parameter of both the engine setting and the operating condition. Depending on the chosen degree of mesh discretization it is possible to fully record the fluid motion and any chemical and thermodynamic phenomena acting in any part of the 3D-CFD-domain up to a length scale which is not far away from molecular dimensions.

But this is the theory. The practice looks quite different. The computational resources in terms of the required hardware and the related CPU-time are prohibitive in case of a too fine mesh discretization. Consequently the 3D-CFD-domain has also to be limited to the part of the engine of particular interest (usually the combustion chamber and parts of the intake and exhaust channels), i.e., this procedure introduces boundary conditions that have to reproduce the part of the engine missing in the 3D-CFD-domain. The setting of the boundary conditions, e.g., due to the difficulty in determining or measuring the flow field in these regions can be very often a source of inaccuracy that irremediably compromises the overall quality of the results and drastically reduces the benefits of such resource investments. Another critical point is represented by few three-dimensional engine process models that due to the lack of phenomena understanding at the fundamental physical level, inaccurate mathematical formulations, numerical dependencies on the mesh structure, ambiguous validation processes, etc., are not able to ensure a high level of reliability in reproducing and predicting the requested engine processes.

As discussed in Chapter 2 the implementation of the 3D-CFD-simulation in the engine development process is a quite controversial topic in which expectations, efforts and resource investments are evaluated in a different manner from developer to developer. Actually, nobody denies the potentiality of 3D-CFD-simulations but, at the end, pragmatically their implementations still remain quite limited in comparison to other simulation tools.

With the development of a new 3D-CFD-tool called *QuickSim* a new approach in the three dimensional analysis of internal combustion engines has been introduced. This tool tries to take advantage of the potentiality of the traditional 3D-CFD-approach combined with a remarkable reduction of the above mentioned drawbacks so that a mayor contribution in engine development process can be ensured. The future reduction of costs for computational resources and their increasing performance will surely support this task. The conceptual ideas that have motivated the development of *QuickSim* are widely reported in Chapter 7.

7

Towards an improved 3D-CFD-Simulation

The interest in enhancing the 3D-CFD-simulation from something for few specialists and with many limitations into a reliable and powerful tool fully integrated into the engine development process is incredibly high. Future engines, turbocharged, downsized and with innovative combustion solutions also for alternative fuels, will require a more detailed analysis of the engine processes in order to better understand the potentialities of the different configurations and then to properly design the engine towards efficiency exploitation. For this task no other simulation tool is more adequate then the 3D-CFD-simulation.

However, any engine manufacturer, as introduced in Chapter 2, seeks simulation tools, whose peculiarities are: fast analysis calculations, reliability, user-friendliness, clear representation of the results without ambiguity and cost efficiency. Moreover a simulation tool must be well integrated into the existent engine development process so that the comparison with both experimental data and other simulation programs can be efficiently set. It can be seen that in the above listed peculiarities the 3D-CFD-simulation needs a lot of improvements.

7.1   An innovative Fast-Response 3D-CFD-Tool: *QuickSim*

The aim of this work is to make a contribution to the mentioned changing-step of the 3D-CFD-simulation by introducing an innovative concept in the simulation of internal combustion engines. This new concept is the basic idea of the 3D-CFD-program *"QuickSim"*, a tool that uses the traditional CFD-code StarCD$^{®}$ from Adapco in background. The programming of *QuickSim* has been started in the year 1998 at the very beginning of my activity as research assistant of Prof. Michael Bargende at both the FKFS (Research Institute for Automotive and Internal Combustion Engines - Stuttgart) and the IVK-University of Stuttgart. In this chapter the main features of *QuickSim* will be analyzed.

7.1.1 **Fast Analysis**

In a 3D-CFD-simulation in each cell of the mesh representing the fluid domain a local numerical solution of partial differential equations for the conservation of mass, momentum, energy and several species concentrations used for describing the mixture formation is performed. In the case of an engine combustion-chamber very often meshes with usually about 500,000 cells, up to 40 "mixture-species-scalars" and very complex and computing time-expensive 3D-models are implemented. All this requires the solution of a "huge" amount of equations that easily leads to an exorbitant CPU-time even using PC-clusters (up to several days).

Figure 7.1 and Figure 7.2 show indicatively (status 2008 using a PC equipped with operating system Linux) the estimated CPU-time for a simulation over a full operating cycle (computation of a mesh with cylinder and parts of intake and exhaust channels over 720 deg) as a function of the total number of cells N_{Cells} in the mesh of the combustion chamber and as a function of its averaged cell-discretization-length l_D, respectively. It is evident that, e.g. halving l_D from 1.5 mm to 0.75 mm, the number of the cells in the engine mesh increases by a factor of eight which causes an increment of CPU-time by approximately a factor of 25 in comparison to the previous one. Even multi-processor computing allows for transient simulations, in the "best case", only a linear reduction of the CPU-time with the number of processors as long as this number is small (number of processors $<\sim 6$).

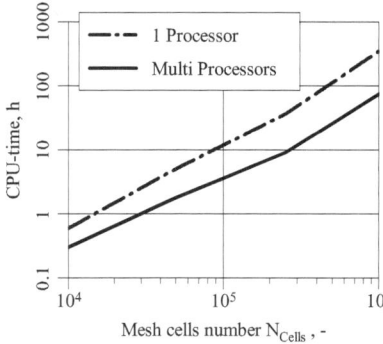
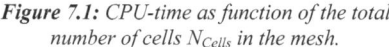
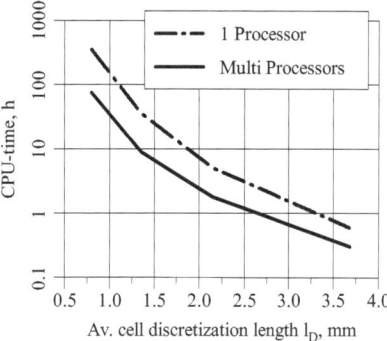

Figure 7.1: *CPU-time as function of the total number of cells N_{Cells} in the mesh.*   **Figure 7.2:** *CPU-time as function of the averaged cell-discretization-length l_D.*

The estimated required time-step Δt in the simulation in order to obtain a temporally accurate numerical solution depends on the convective velocity of the processes $\bar{v}$ in the flow field and the averaged cell-discretization-length l_D of the mesh. Here the Courant number Co introduces the following relation:

$$Co = \frac{\overline{v} \cdot \Delta t}{l_D} \leq 0.7 \tag{7.1}$$

This relation assumes that the chosen time-step Δt must be set low enough to allow each involved cell to "capture" the process transition. Usually, in the simulation of internal combustion engines using meshes with an averaged degree of discretization, setting values of the time-step Δt corresponding to a crank angle step $\Delta\varphi$ of about 0.25 deg are a common practice at middle engine speed. During the exhaust phase, due to high gas velocity in the channel, or generally at low engine speed this value $\Delta\varphi$ is usually reduced to about 0.125 deg, i.e. the simulation of an operating cycle requires from 3,000 up to 6,000 time-steps.

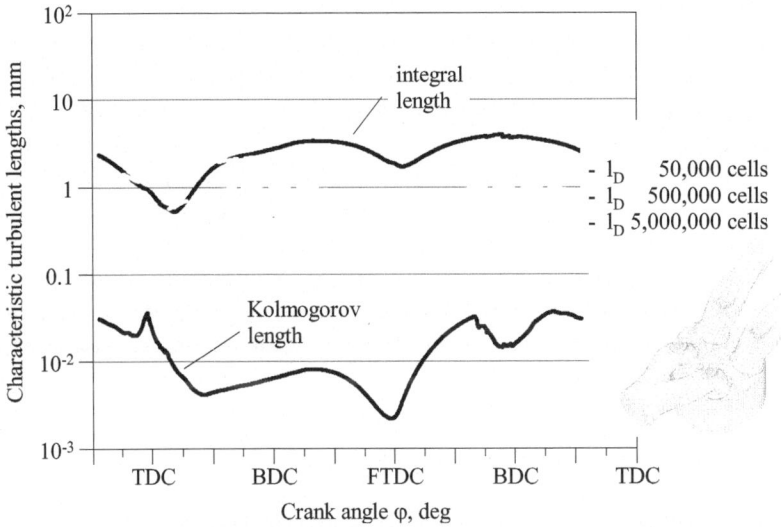

Figure 7.3: Cell discretization l_D and characteristic turbulent lengths (integral and Kolmogorov) – Cylinder displacement: 500 cm³ – 5000 rpm – WOT.

Following the Courant number relation, looking for a more fine mesh e.g. by halving l_D this implies not only more cells and consequently much more computing time for each time-step, but also more time-steps required for the simulation of a cycle. A comparison between the cell discretization l_D and the length scales of the phenomena occurring in an engine helps understanding the capability of both the mesh and the implemented 3D-CFD-models to directly capture the phenomena evolution. In Figure 7.3, on a logarithmic ordinate, the average cell dimensions l_D of the same combustion chamber using three different discretization degrees are reported in comparison to the turbulent length scales l_l and l_k (integral and Kolmogorov turbulent length scale, respectively, that represent the extremes of the turbulent kinetic energy spectrum – see Chapter 6.2.1.3). Here a constant number of cells during the whole calculation is assumed. From this figure it becomes evident that even a mesh with an uncommon high discretization (5,000,000 cells only for the combustion chamber) has an average discretization length l_D much bigger than the scales of many turbulent phenomena. In this case the results of the implemented 3D-CFD-models which work only with the local variables of the flow field solution and do not take explicitly into account the local mesh structure will be mesh-dependent and in many cases absolutely not representative for the phenomena analysis.

7.1.1.1   Mesh Discretization for DNS Simulations

As introduced in Chapter 6.2.1.3 the direct numerical simulation DNS is able to solve any turbulent flow problem with the same procedure as for a laminar flow problem, i.e. without using special "arrangements" for turbulence modeling. To do that, the computational mesh must be extremely refined in order to be able to detect the smallest turbulent eddies, i.e. the dimension of the cells l_D should be many times smaller than the Kolmogorov turbulent length scale l_k (~ 0.015 mm for IC-engines as averaged over the operating cycle – see Figure 7.3). It is evident, that a DNS discretization of a combustion chamber would lead to a total number of cells in the mesh of more than 10^{13} cells and about 6,000,000 time-steps required for the simulation of one operating cycle. This simulation, even finding a computer with enough resources, would then require many years of calculation till a solution. Therefore it cannot be expected to count on the DNS-approach as a development tool for the next future.

7.1.1.2   Mesh Discretization for LES Simulations

Large-eddy simulations LES mean the solution of a turbulent flow problem using a direct numerical simulation DNS except for the turbulent eddies that have a length scale l smaller than a reference length scale l_s ($l_k < l_s < l_l$, for more details see Figure 6.4). To do that, the computational mesh must be enough refined to be able to detect the turbulent eddies solved by the DNS (large scale eddies) i.e. the dimension of the cells l_D should be many times smaller than the reference length scale l_s. The unresolved turbulent scales are modeled as a isotropic turbulence using turbulence models, such as the $\tilde{k} - \tilde{\varepsilon}$-model. This kind of approach is much more efficient then pure DNS simulations, but still meshes with many millions of cells are required, what still demands a prohibitive amount of CPU-time. Therefore the practical interest in LES-simulations for the investigation of internal combustion engines is still very low.

7.1.1.3 Mesh Discretization for *QuickSim* Simulations

The common and traditional approach in the 3D-CFD-simulation is usually based on the implementation of isotropic turbulence models for the solution of all turbulent eddies independent from their length scale l. Here the computational mesh does not require a "strict" grid refinement, so that it is possible to use more coarse meshes than in the case of DNS- and LES-simulations. As introduced before, usually meshes for the combustion chamber with approx. 500,000 cells are implemented in the calculation and during the years the tendency has been to continuously increase the cell refinement (up to 1,000,000 cells).

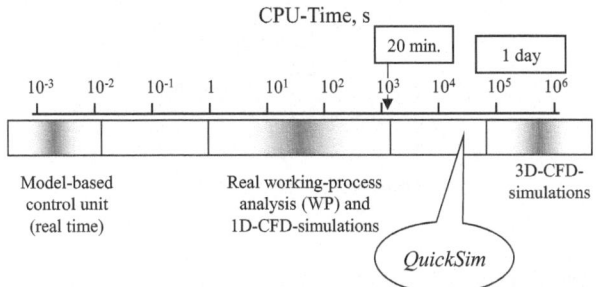

Figure 7.4: *The 3D-CFD-tool "QuickSim" in the spectrum of CPU-time among different calculation/simulation tools.*

In comparison to a traditional 3D-CFD-simulation that is able to perform a numerical analysis of any fluid-dynamical object, the 3D-CFD-tool *QuickSim* is dedicated only to the simulation of internal combustion engines. In combination with the implementation of improved or newly developed 3D-models for a reliable formulation of only engine processes, which take also the cell dimension and mesh structure into account, this introduces a considerable reduction of the number of cells in the mesh. Thanks to this approach *QuickSim* allows a reliable simulation using meshes with about 30,000 ÷ 50,000 cells for the discretization of the combustion chamber depending on the design complexity. The averaged crank angle step is approx. 0.5 deg so that the calculation of one operating cycle requires ca. 1,500 steps. In addition the mixture composition of the working fluid is described using only six "mixture-species-scalars" (see Chapter 8.1). At the end, the sum of these actions allows a CPU-time reduction up to a factor 100 in comparison to the traditional approach, only by an increasing of the averaged discretization length of the cells up to approximately a factor 2.5. Figure 7.4 shows that *QuickSim* finds a new location in the spectrum of CPU-time among different simulation tools, i.e. it permits to perform a wide range of simulations with high reliability and predictability that would be unacceptably too time expensive using a traditional 3D-CFD-simulation.

7.1.2   Reliable Calculation

As introduced in the previous paragraph the 3D-models implemented into *QuickSim* do not have a general validity for any thermodynamic investigation. Their formulation is trimmed in order to both optimize the solution of engine processes and reduce the computational resources required for the calculation.

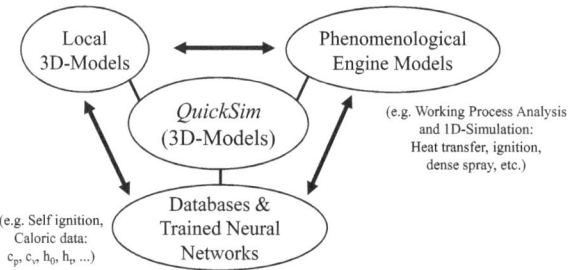

Figure 7.5: *General schematic of 3D-engine-models in "QuickSim".*

These models are based on a combination of different approaches (traditional local 3D-CFD-models, engine-specific phenomenological relationships, trained neural networks, databases and if needed empirical relationships – see Figure 7.5). In the development of the models the cell dimensions and cell structures have been taken explicitly into account. This combination allows a reliable analysis of the processes whose behavior is relevant for practical applications in the engine design process; namely the thermodynamic aspect (see Chapters 8, 9 and 10). E.g. with focus on 3D-modeling of the combustion process, it is more desirable - and absolutely satisfactory at the moment - to have a time-convenient and accurate prediction of the fuel heat-release and a realistic flame propagation calculation instead of a time-expensive detailed analysis of the chemistry involved in the process. In *QuickSim* an internal coupling between the 3D-CFD-calculation and the real working-process analysis (WP) allows to compare the "global" results (heat release, wall heat transfer, internal energy variations, etc.) at each time-step so that any implausible differences can be immediately recognized and eventually corrected.

7.1.3 User-Friendliness

In the traditional 3D-CFD-simulation the setting of the mesh motion, boundary conditions, initial conditions, definition of the mixture formation, activation – if available at all - of all necessary 3D-models (e.g. thermodynamic properties of the working fluid, ignition, combustion, injection spray and wall heat transfer) is a very long and laboriously procedure related to a complex and time expensive control and debugging steps until the simulation can be properly started.

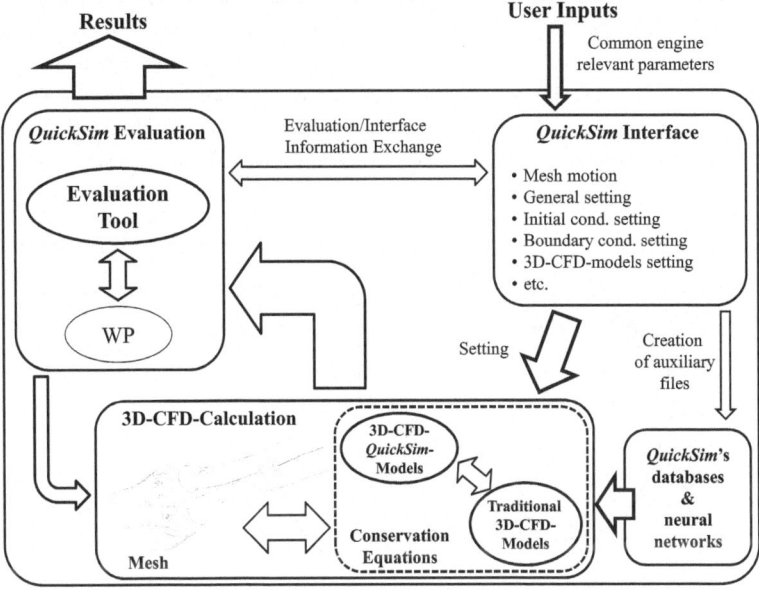

Figure 7.6: The interface tool within the layout of "QuickSim".

The 3D-CFD-code *QuickSim* as an engine dedicated-tool provides an efficient interface for the setting procedure. Similar to other dedicated tools (e.g. GT-Power® by Gamma Technologies for the 1D-CFD-simulation and Tiger by EnginOS for the real working-process analysis) this procedure is standardized, reduced to a minimal number of inputs for a clear and simple definition of the engine and the operating conditions that has to be investigated. Here, it is not required that the user has an "exceptional experience" in 3D-CFD-simulations, because in this procedure, he deals with more familiarly parameters commonly used in the engine development process. It is the duty of *QuickSim* to convert these parameters instantly and accurately into a traditional setting required by the simulation. Finally the interface tool is able to provide automatically all necessary auxiliary files (databases and trained neural networks) that are required by the 3D-CFD-engine models (see Figure 7.5).

7.1.4   Clear Representation of the Results

Results from the 3D-CFD-simulation, mainly as colored flow field maps, are for sure very impressive and allow a detailed visualization of the engine processes. An interpretation from such visualizations towards an objective evaluation of the engine behavior or, e.g. a quantification of the potentiality that can be exploited from a particular engine solution would be in many cases misleading. In addition, any comparison or validation of these results with other simulation tools or experimental investigations is actually not possible, so that at the end results from 3D-CFD-simulations are not of a main interest in the decisive steps of the engine design process.

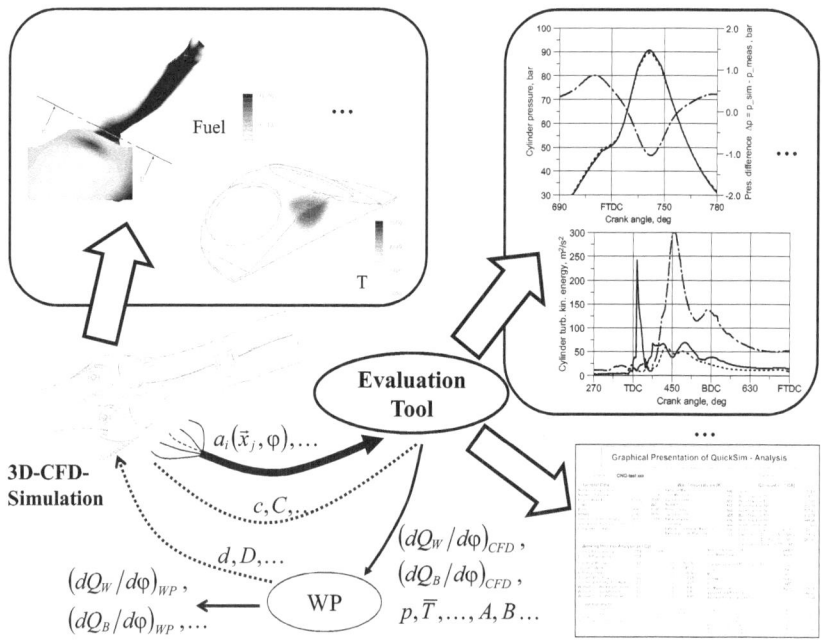

Figure 7.7: *3D-CFD-Simulation with "QuickSim"- A virtual test bench.*

Therefore a comprehensive evaluation tool has been implemented into *QuickSim*. This tool, among other things, collects, averages and extrapolates the countless variables $a_i(\bar{x}_i, \varphi)$ (temperature, pressure, velocity, species concentration, etc.) provided by the 3D-CFD-simulation for each cell i in the mesh at any time-step or crank angle φ (see Figure 7.7). Parts of the outputs of the evaluation tool are similar to the output of a test bench with a modern indicating system integrated with a real working-process analysis, other e.g., are similar to the results provided by LIF-technologies in a pressure chamber for the investigation of complex phenomena like those occurring during the fuel injection (analysis of the spray penetration, droplet size, droplet velocity, etc.).

By means of this procedure the evaluation tool of *QuickSim* provides results which are familiar to any engineer involved in the engine design process. In particular the integration or "internal coupling" between the 3D-CFD-simulation and the real working-process analysis (WP) allows a better comparison and control with test bench results and other simulation tools. The term "virtual test bench", as explained in Chapter 2, reported in Figure 7.7 does not mean a stand-alone engine development without experimental support. It means more the capability of the tool to act as a "virtual eye" inside the test bench, permitting to visualize, measure and record also processes that are hardly to be analyzed – if at all - at the experimental test bench (e.g. mixture formation, fluid motion, turbulence, residual gas distribution, etc.).

7.1.5   Cost Efficiency

Thanks to the reduced number of cells and an efficient implementation of engine process models, 3D-CFD-simulations with *QuickSim* do not require expensive computational resources. In comparison to traditional 3D-CFD-simulations, the simulations run on common PCs within a very impressive computational time (using a single processor just few hours for one operating cycle for the simulation of one cylinder with intake and exhaust manifolds). In the last years *QuickSim's* capability of providing results within short time has permitted to accompany and support the development of several engines. Also in Motorsports [44,62,63,64] where the development timetable is especially tight the tool has found a large application.

7.1.5.1   Processor Utilization for *QuickSim* Simulations

PC-cluster solutions are required in order to shorten the CPU-time of a calculation to an acceptable level, but especially in case of transient simulations this is not a very efficient approach. Here the entire 3D-CFD-mesh is divided into many "sub-domains", each one assigned to one processor of the cluster. During the simulation, at any corrector-step [56], each processor must spend part of its calculating time in exchanging data with the other processors so that all the sub-domains reciprocally and correctly pass the "internal" boundary conditions to their neighbors. By increasing the number of sub-domains this request of data exchange increases disproportionately to the effective CPU-resources for flow field solution, therefore at a certain number of processors in the cluster the efficiency of the simulation drastically decreases.

Considering the performance of *QuickSim* in terms of need of computational resources, instead of using complex systems of PC-clusters, it is more convenient to run each simulation only on one processor and then use the total amount of processors available for the simultaneously running of different variants. First of all, this permits a more efficient utilization of the available processors and then it remarkably simplifies the setting and control of the simulation, which is highly time expensive in case of PC-clusters.

7.2   Additional Features of *QuickSim*

In this paragraph additional innovative features of *QuickSim* that has been realized during the last years will be introduced (more details in Chapter 11).

7.2.1   Simulation of several successive Engine Operating Cycles

The tool *QuickSim*, thanks to the convenient CPU time, has been developed in order to allow to simulate successive operating cycles instead of the common practice whereas the calculation starts before the opening of the intake valve (IVO) and terminates, if at all, after the end of the combustion. In order to perform this, first of all, the difficulty of correctly solving the flow through the exhaust valve ports has to be managed. Here the fluid velocity is extremely high and in case of high pressure differences between combustion chamber and exhaust channel critical conditions (Mach number >1) are given. The second main difficulty is the resetting of all evaluation parameters at the end of the cycle and in particular the redefinition of the species describing the mixture in order to allow a correct simulation of the successive combustions. E.g. a distinction of the exhaust gas between residual gas EGR and newly produced combustion products must be adequately performed otherwise it is not possible to accurately calculate the flame propagation, that directly depends on the concentration of EGR in front of the flame (unburned zone).

The capability of *QuickSim* to perform successive operating cycles allows to remove the relevant influence of the fluid initial conditions in the mesh on the results. The 3D-CFD-simulation of an internal combustion engine is an unsteady-flow calculation (transient simulation) [53,54]. Starting from initial conditions of the flow variables at t_0 or φ_0 (see Figure 7.8) the solution of the flow variables in the conservation equations for each successive allowable size of computational time-step Δt is performed by iterative algorithms based on the predictor-corrector-strategy (e.g. PISO algorithm [56]).

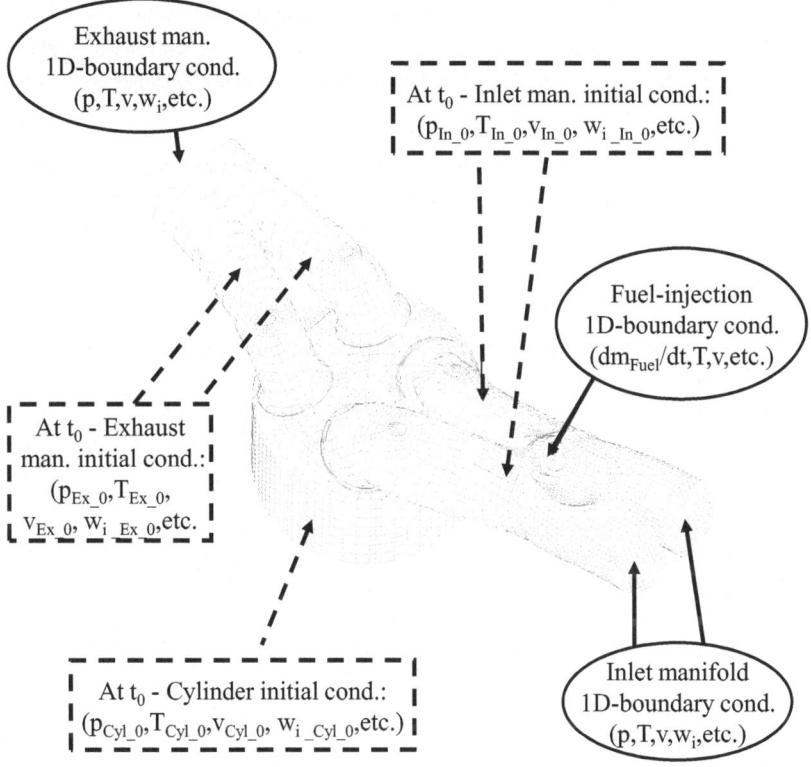

Figure 7.8: *3D-CFD mesh – Initial and boundary conditions.*

The setting of the initial conditions requires high accuracy; otherwise the results of the following time-steps will not represent the real flow field correctly. Since it is practically impossible to know the flow field "a priori" at the beginning of the simulation (φ_0 is usually before IVO) it is necessary to make some simplifications. Generally the 3D-mesh is divided into three regions (cylinder, inlet and exhaust manifold, respectively). For each region, usually, no fluid motion, and an estimated homogeneous distribution of the following flow variables over the volume are assumed: pressure (i.e. no pressure waves), temperature, fluid composition etc. Therefore this is no more than a "rough" estimation of the flow field at the start of the simulation. Depending, first of all, on the operating condition of the engine (rpm, load, etc.) and then by the dimension of the 3D-CFD-mesh (see Figure 7.8, Figure 7.9 and Figure 7.10), some time from the simulation start φ_0 is required in order to establish a realistic flow field within the 3D-CFD-mesh. This is called the "running-in phase". Very often this phase can be assumed terminated when the starting fluid in the 3D-CFD-mesh defined by the initial conditions has been entirely removed and replaced by the new one inserted by the boundary-conditions.

In the Chapter 11.8 some results will show that the first simulated operating cycle is not representative for the engine behavior. Sometimes even more than ten cycles are necessary in order to achieve a satisfactory convergence of the engine results ($imep$, $p_{IP/SOI}$, p_{max}, combustion duration, etc.).

7.2.2   Extension of the 3D-CFD-Domain up to a Full-Engine Simulation

During the years, the successive development steps of *QuickSim* led to the possibility to extend the 3D-CFD-domain (i.e. the extension of the cylinder mesh to other regions of the engine) without expecting an exorbitant increasing of the CPU-time.

Simulation of the Cylinder

At the beginning of the tool development the discretized 3D-CFD-domain was limited only to the combustion chamber of one cylinder with parts of the intake and exhaust manifolds (usually up to the location of the pressure sensors, see Figure 7.8). Because of the discretization of a part of the engine, boundary conditions at the fluid-dynamical borders of the 3D-mesh are required (i.e. at the 3D-cuts of the exhaust and inlet manifolds, respectively, at the fuel injector nozzle and eventually at the piston crevice as blow-by flow).

Actually the boundary conditions describe the time-resolved fluid-dynamical behavior of the missing parts of the engine. Focusing now on the boundary conditions at the manifolds these are usually provided partly by measurements (e.g. pressure traces from a low-pressure-indicating-system in the manifolds) and additionally or entirely by 1D-flow-models (e.g. GT-Power). Since it is practically impossible to set boundary conditions locally resolved over the border regions, they are assumed homogeneous (1D-time-resolved boundary conditions – see Figure 7.8), so that a switch 1D-to-3D-flow-field occurs.

The locations of these necessary switches, which "irremediably" influence the flow field, must be also chosen properly; otherwise they can be a source of grave inaccuracies (see e.g. Chapters 11.5 and 11.6). The best locations are where the pipes are straight with a constant cross section and in which mixing processes between air, fuel or exhaust gas do not take place. For the above mentioned reasons, especially the optimal length of the discretized part of the intake system $L_{3D-Mesh}$ must be chosen accurately (see Figures. 7.7 and 7.8), e.g. a short $L_{3D-Mesh}$ decreases both the CPU-time (due to a reduced number of 3D-cells in the mesh) and the "running-in phase" (see Chapter 7.2.2), but on the other hand, in case of a fuel injection into the inlet manifold or into the airbox or where backflows from the cylinder are expected, leads to an inappropriate location of the 1D-to-3D-flow-field switch. In this work, it will be shown, that in the most cases it is better to locate the 1D-to-3D-flow-field switches in appropriate locations and to invest more both for the "running-in phase" and for the CPU-time of each operating cycle.

Integrated Simulation of the Cylinder with the Airbox

Since 2002 accurate investigations have been performed with meshes that include one cylinder and the whole airbox (see Figure 7.9). In this case there are no manifold 1D-boundary-conditions between the airbox and the simulated cylinder, i.e. the charge flows "undisturbed" in a 3D-field from the throttle to the cylinder, and both a backflow into the intake manifold and the fuel injection can be reproduced with accuracy.

For the other cylinders in *QuickSim* (e.g. 1 and 3 in Figure 7.9) boundary conditions must be provided either by a direct import of pressure traces from the sensors or by 1D-CFD-programs. Limitations in this approach are mainly represented by very strong backflows of residual gas or injected fuel into the airbox, where contributions of the missing 3D-CFD-cylinders could be wrongly predicted by the manifold 1D-boundary-conditions.

Because, as well known, 1D-CFD-simulation programs are not able to accurately reproduce the behavior of the airbox in details, these kind of simulations with *QuickSim* have permitted to investigate the influence of the airbox design on the exchange process, the mixture formation and the combustion of the reference cylinder, i.e. in comparison to the simulation of the cylinder alone both the accuracy of the results and the predictability can be remarkably increased. Starting from an initial design of the airbox and the combustion chambers it is possible to virtually investigate design variants up to a certain degree of modification, at least so far that reasonable 1D-boundary-conditions for each variant can still be provided.

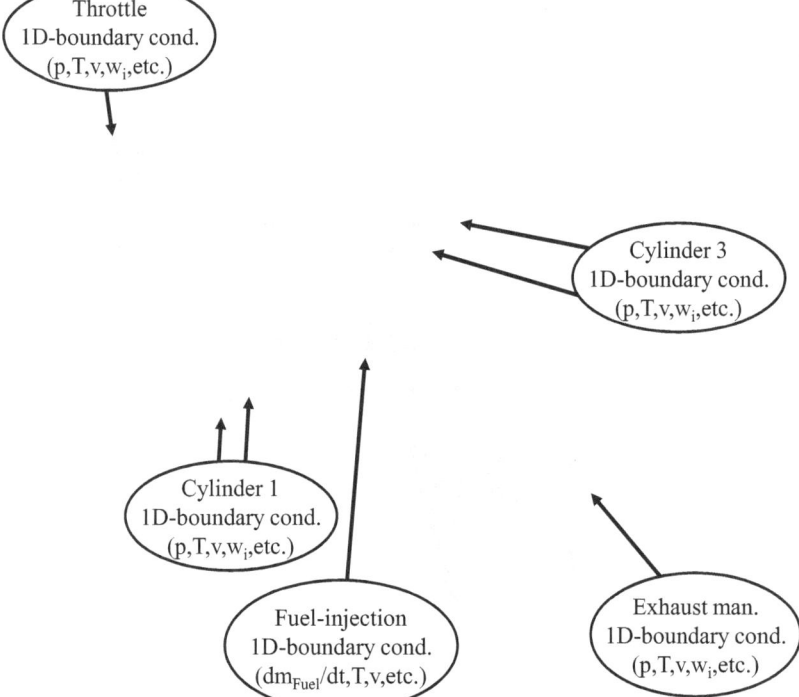

Figure 7.9: *"QuickSim" – Cylinder with airbox simulation.*
(3 cylinders turbocharged CNG engine).

Full Engine Simulation

A recent development step of *QuickSim* (November 2009) has permitted to extend the 3D-domain to the full engine; reasonably intended as the domain that includes the airbox, all the cylinders and the exhaust system up to the inlet of the turbine (see Figure 7.10). Here between the throttle and the cylinders there is no manifold 1D-boundary-condition, i.e. the charge flows "undisturbed" in a 3D-CFD-field through the whole airbox and all the cylinders. This allows, among other things, a detailed analysis of the differences among the cylinders with focus on volumetric efficiency, residual gas concentration at IVC, fuel mixture formation, turbulence profile up to the end of combustion, combustion progression etc.

A full engine simulation with *QuickSim* ensures a very high level of predictability, that allows to investigate the influence of design modifications, different injection strategies and valve timings on the performance of the "entire" engine (a decisive step towards virtual engine development). In contrast to simulations of only one cylinder where it has to be assumed or at least supposed that the simulated cylinder follows the behavior of the entire engine, full engine simulations permit to analyze and evaluate the engine behavior in its completeness.

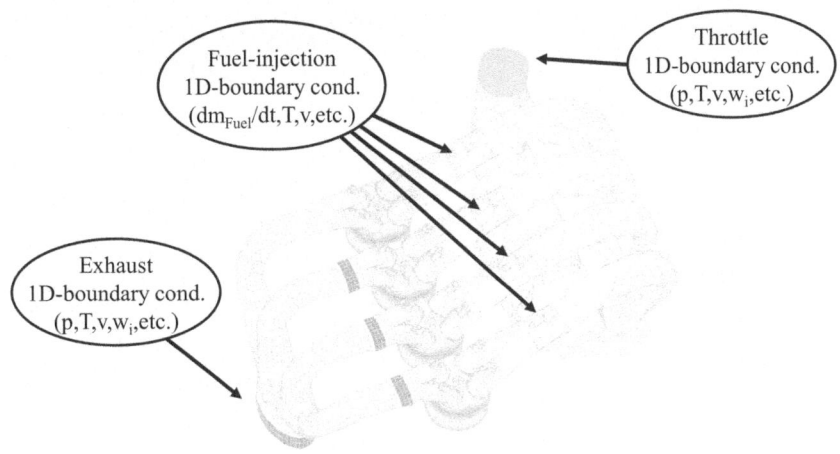

Figure 7.10: *"QuickSim" – Full engine simulation*
(4 cylinders gasoline engine).

In Chapter 11 a detailed analysis of *QuickSim*-results of the same engine with meshes having a different extension (from the cylinder alone up to the full engine) will be reported. This analysis focuses on the influence of both the initial conditions (simulation of successive operating cycles) and the location of the boundary conditions on the simulation results in comparison to experimental data at the test bench.

7.2.3   The Simulation of a Flow Test-Bench

An interesting feature of *QuickSim* is represented by the flow test-bench simulation (see Figure 7.11). Starting from the mesh of the combustion chamber and parts of the intake and exhaust manifolds, respectively, used for the common engine simulations, the program is able to automatically undertake the necessary modifications in order to reproduce the device at the flow test bench. The boundary conditions at the mesh with the required pressure difference (usually Δp =5000 Pa) between cylinder and manifolds are set automatically. The simulation is performed under transient conditions so that during one run it is possible to determine the discharge coefficients of the valves with all relevant lifts.

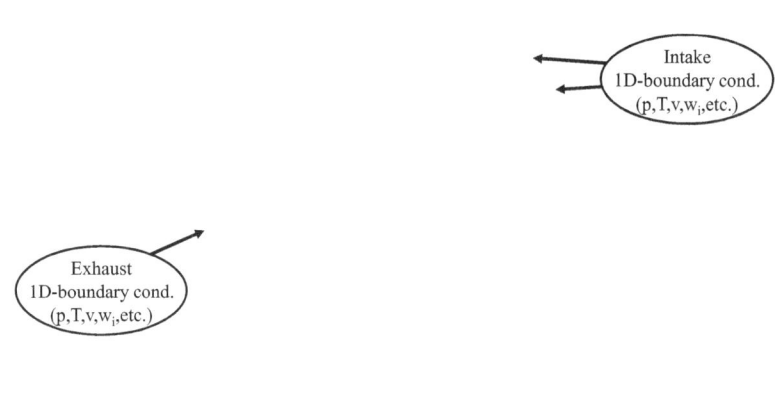

Figure 7.11: *"QuickSim"- Virtual flow-test-bench.*

The simulation starts with closed valves and then each valve opens slowly up to the first relevant valve-lift. At this point the motion of the valves is frozen until the flow through the combustion chamber has reached stationary conditions while the evaluation tool of *QuickSim* records and calculates all necessary information, among others also the discharge coefficient. Successively and continuously the mesh slowly moves to the next relevant valve lifts by repeating always the same evaluation procedure up to the end of the measurement schedule.

7.3   **Summary of the *QuickSim* Features**

The following summary wants to collect and underline the main features of *QuickSim*. This serves mainly as an outlook for the descriptions in the next chapters and for future improvements.

1. *QuickSim* is a 3D-CFD-tool dedicated and optimized for the simulation of internal combustion engines (gasoline, diesel, CNG and other alternative fuels). There are no limitations regarding fuel injection solutions and valve motions.

2. *QuickSim* is a fast response simulation tool that thanks to a reduction of mesh discretization allows very competitive CPU-times (up to a factor 100 in comparison to traditional 3D-CFD-simulations).

3. Efficient utilization of the processors. Contemporaneous simulation of different engine variants or operating conditions instead of complex PC-cluster solutions.

4. Improved or newly developed 3D-models for the description of engine processes that ensure an efficient and reliable calculation also using coarse meshes.

5. An integrated and automatic "evaluation tool" for a comprehensive analysis of the relevant engine parameters (clear representation of the results).

6. Integrated 3D-CFD-simulation and real working-process analysis (WP) for both a supported analysis of the engine processes and a better comparison and control with test bench results and other simulation tools.

7. Easy and fast simulation setting.

8. Easy and fast simulation control.

9. Simulation of successive operating cycles (reduction up to completely elimination of the influence of the initial conditions).

10. Extension of the simulated 3D-CFD-domain up to the full engine (increasing of predictability and reduction of the influence of boundary conditions).

11. Simulation of a virtual flow test bench.

7.4   *QuickSim*'s Calculation Layout

Before introducing a selection of 3D-engine-process models in the next chapters, the basic idea of the calculation layout of *QuickSim* is reported here (see Figure 7.12). This layout is based on a modular structure of engine models, databases or neural networks, variable/parameter exchange devices and control functions [21,22,63]. The information/parameter can be assigned to the following three main groups:

Test bench and laboratory environment

Here, if available, measurement data (or eventually results from extern simulation tools) is provided for automatic calibration, comparison and validation.

Zero-dimensional environment

The core of this calculation environment is the evaluation tool that collects, averages, extrapolates, etc., the countless variables $a_i(\vec{x}_j, \varphi)$ (temperature, pressure, velocity, species concentration, etc.) which are provided by the 3D-CFD-simulation for each cell j in the mesh at any time step or crank angle φ.

Parts of the outputs of the evaluation tool are similar to the output of a test bench with a modern indicating system, others e.g., are similar to the results provided by LIF-technologies in a pressure chamber for the investigation of complex phenomena like those occurring during the fuel injection (spray penetration, droplet size, droplet velocity, etc.).

In a second step the outputs of the evaluation tool are progressively at disposal for other kinds of applications like the real working-process analysis WP (zero-dimensional or thermodynamic analysis of the engine operating cycle). In this case the real working-process analysis uses the 3D-CFD-simulation as a virtual engine test bench. Another implementation of the evaluation tool consists of establishing a real time feedback of any variable as an enhanced information exchange process between the evaluation process and the 3D-CFD-simulation. This approach allows the 3D-CFD-simulation to have both at disposal, local and global variables in each cell of the mesh. The implementation of both variable types in the 3D-models can be properly combined (e.g. when a local variable due to convergence difficulties is not reliable anymore or when a formulation of a phenomenological or quasi-dimensional model is recommendable).

In a third and last step all the outputs from the WP are progressively at disposal as a feedback for the 3D-CFD-simulation for a comparison between the two approaches and eventually for control.

In this way it is possible to establish an internal coupling between the 3D-CFD-simulation and the WP at any time step, both based on the same simulation evolution.

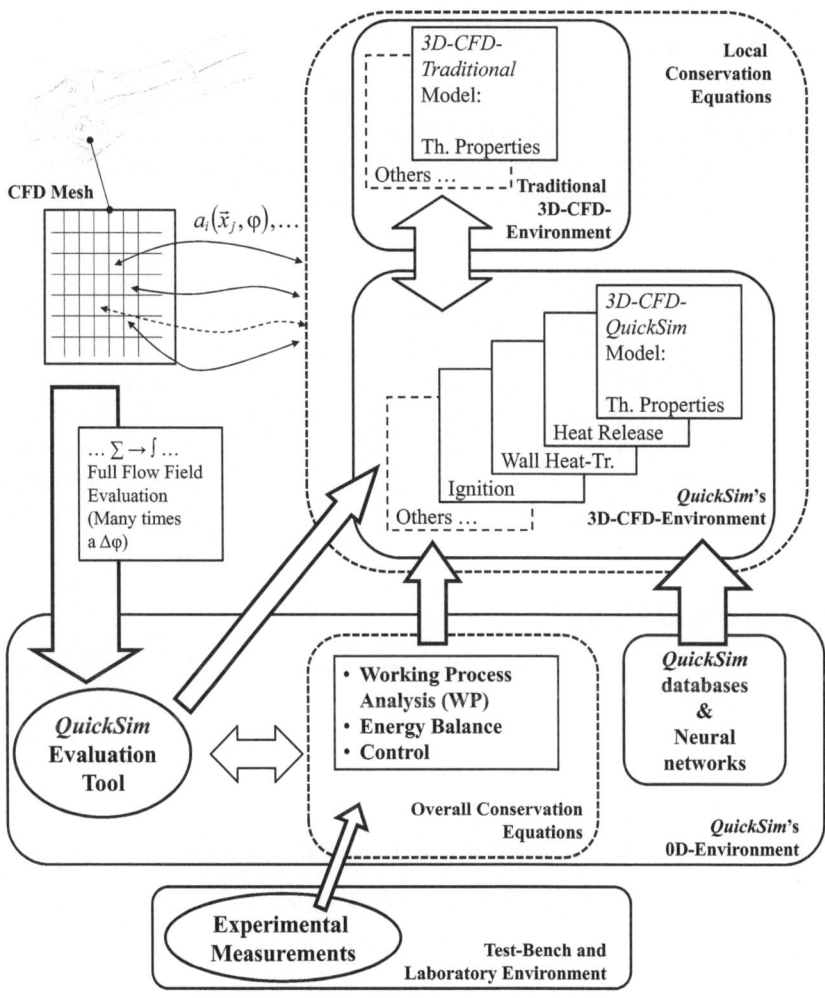

Figure 7.12: Calculation layout in "QuickSim".

The last element of this environment in *QuickSim* is represented by a collection of libraries (databases and neural networks) which help the local 3D-CFD-engine-models with fast response inputs whose direct calculation would be too time expensive (e.g. the thermodynamic properties of the working fluid, ignition delay, etc.)

3D-CFD-Environment

The background of this environment is the solution of the conservation equations solved by the 3D-CFD-code StarCD. As mentioned in Chapter 2.3 the conservation equations are the "webs" for all the information transfers among the engine process models that are needed for calculating the final results ensuring the correct mass, momentum and energy balance at the local level.

Few phenomena or engine processes are calculated using traditional 3D-CFD-models (e.g. turbulence, viscosity, spray collision, droplet vaporization, etc.), others have been replaced by three-dimensional *QuickSim*-models. The formulation of these models can be conveniently chosen regarding the peculiarities of the investigated process (physical understanding, mathematical and algebraic description), the numerical solution (in particular the mesh dependence) and the required computational time. The information exchange within the calculation layout of *QuickSim* is always able to provide the necessary inputs also for "unconventional" formulations of local 3D-CFD-models [21,22,23,60,63,65,66].

New developed 3D-CFD-Models

Here a selection of newly developed 3D-CFD-models implemented into *QuickSim* in the last ten years is reported:

- model for the thermo-physical properties of the working fluid for any fuel $C_nH_mO_rN_q$
- wall heat transfer model
- spark ignition model (flame propagation model in the near region of the spark plug)
- heat release models for spark ignition and diesel engines
- self-ignition model
- model for gas and liquid fuel injection [67,68]
- model for the dense spray region

- model for choking flows (Mach ≥ 1)

- model for flow streams through the valve seats at low valve lift

- model for an improved setting of initial conditions

- model for an improved setting of boundary conditions

8

3D-CFD-Modeling of the Thermodynamic Properties of the Working Fluid

The solution of the operating cycle of an internal combustion engine - in comparison to thermodynamic machines with external fuel combustion or heat generation (steam engines, Stirling engines, etc.) that work with a fluid (water, helium, etc.) with well defined thermodynamic properties - becomes more complicated due to the determination of the changing properties of the working fluid as a result of internal chemical reactions.

As introduced in Chapter 4 the thermodynamic properties of the working fluid are fundamental terms in the conservation equations and the reliability of the results (especially at high temperatures) is drastically influenced by the description of these properties. Since the thirties, this topic has been investigated intensively and, during the years, many approaches for a adequate description of the working fluid of internal combustion engines have been presented [12,27,28,30,34,35]. These approaches spread from simple empirical formulations up to complex solutions of chemical reaction schemes.

8.1 Introduction

The main target of both the 3D-CFD-simulation and the real working-process analysis is the solution of the engine operating cycle with focus on the "thermodynamic aspect". For this purpose only the following data are required:

- R: the real gas constant of the mixture ($R = \Re / M$)

- h_{tc}: the total enthalpy of the mixture ($h_{tc} = h + h_f$: thermal and chemical term)

The determination of the chemical composition of the working fluid is not a mandatory task, i.e. a reliable empiric formulation for the calculation of the above mentioned properties would be absolutely sufficient. In reality, as intensively reported in the literature, this kind of formulation represents no more than a coarse approximation suitable just for standard fuels and is not able to deliver reliable results for a "state-of-the-art" investigation. Hence, accurate information about the chemical composition of the working fluid is necessary for a more valid approach.

8.2 Chemical Composition of the Working Fluid Mixture

The working fluid in an internal combustion engine is composed by air, fuel and exhaust products (burned gas). The chemical composition and the resulting thermodynamic properties of air and fuel can be easily determined, but in case of the exhaust products the difficulty drastically rises as soon as the molecular structure of the used fuel differs from hydrogen.

8.2.1 One-Step Fuel-Oxidation Reaction Mechanism

The reaction scheme presented here describes a simple oxidation process for an arbitrary fuel $C_nH_mO_rN_q$. Starting from the combustion reactants (fresh mixture of fuel and air) through the oxidation process, the final combustion products can be directly determined (intermediate combustion products are explicitly not taken into account). Depending on the amount of oxygen available for the oxidation process the starting mixture can be classified into three categories [5]:

Stoichiometric mixture

The stoichiometric reaction describes the fuel oxidation process with the minimal oxygen amount necessary for a complete fuel oxidation. In this case O_2 is no longer present in the reaction products and it is assumed that nitrogen contained in the air is not involved in any reaction and the nitrogen eventually contained in the fuel is recombined into N_2. Using the element balance the relation becomes:

$$C_nH_mO_rN_q + \left(n + \frac{m}{4} - \frac{r}{2}\right)(O_2 + 3.773 \cdot N_2) =$$
$$nCO_2 + \frac{m}{2}H_2O + \left[\frac{q}{2} + 3.773 \cdot \left(n + \frac{m}{4} - \frac{r}{2}\right)\right]N_2$$

(8.1)

Lean mixture

With more than the stoichiometric air requirement it is assumed that the oxygen in excess is not involved in the reaction and similarly to nitrogen, the oxygen eventually contained in the fuel is recombined into O_2. Using the element balance the relation becomes:

$$C_nH_mO_rN_q + \lambda \cdot \left(n + \frac{m}{4} - \frac{r}{2} \right) \cdot \left(O_2 + 3.773 \cdot N_2 \right) =$$
$$nCO_2 + \frac{m}{2}H_2O + (\lambda - 1) \cdot \left(n + \frac{m}{4} - \frac{r}{2} \right) \cdot O_2 + \left[\frac{q}{2} + \lambda \cdot 3.773 \cdot \left(n + \frac{m}{4} - \frac{r}{2} \right) \right] N_2 \qquad (8.2)$$

Rich mixture

With less than the stoichiometric air requirement the fuel cannot be fully oxidized. In addition to the common combustion products CO_2 and H_2O also CO and H_2 are present in the final combustion products:

$$C_nH_mO_rN_q + \lambda \left(n + \frac{m}{4} - \frac{r}{2} \right)\left(O_2 + 3.773 \cdot N_2 \right) =$$
$$n_{CO_2}CO_2 + n_{CO}CO + n_{H_2O}H_2O + n_{H_2}H_2 + \left[\frac{q}{2} + \lambda \cdot 3.773 \cdot \left(n + \frac{m}{4} - \frac{r}{2} \right) \right] N_2 \qquad (8.3)$$

The determination of the composition of rich mixtures is remarkably more complex than in the previous cases because the solution cannot be calculated from an element balance alone. In this case an additional assumption about the chemical composition of the product species must be made. The following, often proposed reaction (called the water-gas reaction [5]) sets a chemical equilibrium among the principal product species at common temperatures for combustion processes ($T > 1700$ K):

$$CO_2 + H_2 \Leftrightarrow CO + H_2O \qquad (8.4)$$

The equilibrium constant K_{1_Step} of this reaction (see more details in Chapter 8.4.1.1), that links the mole concentrations of CO_2, H_2, CO and H_2O, becomes:

$$K_{1\_Step} = \frac{n_{CO} \cdot n_{H_2O}}{n_{CO_2} \cdot n_{H_2}} . \qquad (8.5)$$

As usual in numerical applications this equilibrium constant will be described with a polynomial function (e.g. coefficients from JANAF-tables [57]):

$$\ln K_{1\_Step}(T) = \sum_{j} \frac{a_j}{T^{j-1}} = 2.743 - \frac{1761}{T} - \frac{1611 \cdot 10^3}{T^2} + \frac{280.3 \cdot 10^6}{T^3} \qquad (8.6)$$

thus at the end, in order to unequivocally determine the composition of all the species, a temperature T for the reaction equilibrium must be assumed.

Resulting Mixture Composition

In Figure 8.1 with focus on rich mixtures the variations in the composition of the exhaust gas (e.g. combustion products from the reaction between iso-octane and air) are reported.

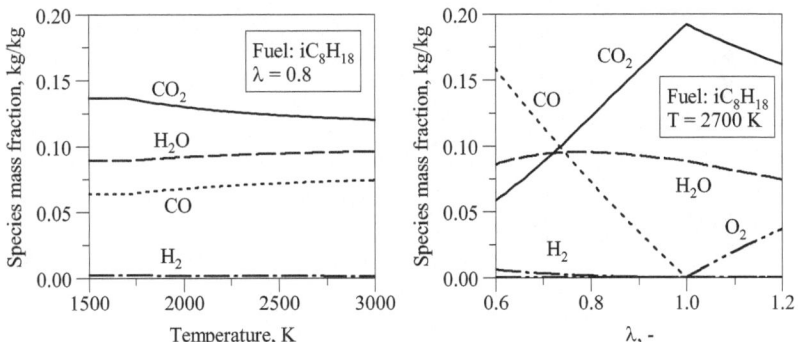

Figure 8.1: *One-step fluid oxidation: composition variation of CO_2, H_2, CO, H_2O and H_2 as function of temperature and λ (fuel: iso-octane).*

The approach described before is very simple and considers the burned gas composed only by five species: CO_2, H_2O, CO, H_2 and N_2. However more critical than the limited number of species considered in the reaction mechanism is the disregard of essential processes at high temperature (e.g. the dissociation effect of CO_2) that remarkably influence the composition of the burned gas (see 8.4.1).

8.2.2 The Reality: More than Thousand Intermediate Products

The combustion process of an arbitrary $C_nH_mO_rN_q$ fuel looks quite different then the mechanism presented in the one-step fuel-oxidation approach (see Chapter 8.2.1). In Table 8.1 the number of chemical reactions and species involved in the combustion mechanism of different fuels is reported. It becomes evident how the complexity of the reaction scheme remarkably increases with rising complexity of the molecular structure of the implemented fuel. In particular, the combustion process at low temperature (LTC), due to the very low reaction speed, drastically increases the number of intermediate products (especially radical species) involved in the combustion process. E.g. the combustion process of a common gasoline fuel at low temperature can be described with approximately 1000 chemical reactions and more the 7000 species.

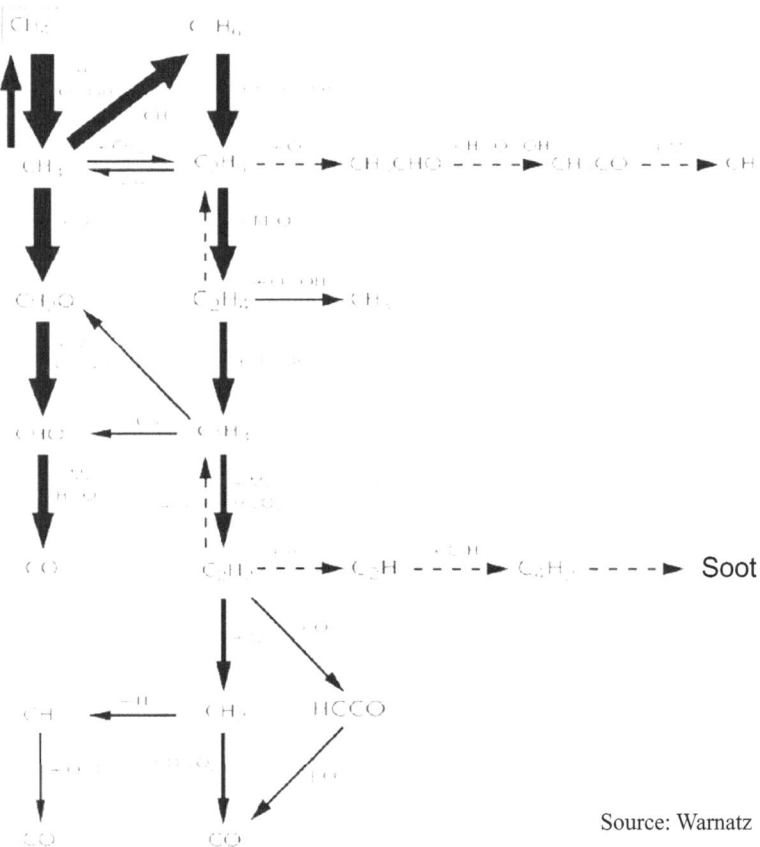

Figure 8.2: *The reaction scheme (CH₄ oxidation with O₂ at high temperature).*

Figure 8.2 shows exemplarily the reaction scheme of CH₄ with pure oxygen at high temperature. Also this simple case shows a considerable number of reactions and species to be solved during the 3D-CFD-simulation of this combustion process. Since, as discussed in Chapters 6 and 7.1.1, any species explicitly considered required an additional conservation equation, which remarkably increases the CPU-time, a detailed and comprehensive analysis of the reaction scheme of common fuels is actually not affordable for practical 3D-CFD-simulations.

Table 8.1: Combustion of different fuels: complexity estimation.

Fuel	No. of Species	No. of Chemical Reactions
H_2 / O_2 Oxidation	8	40
CH_4 / O_2 Oxidation	34	400
CH_4 / O_2 / N_2 Oxidation	54	640
n-C_7H_{16} / i-C_8H_{18} / O_2 Oxidation	100	740
n-C_7H_{16} / i-C_8H_{18} / O_2 Oxidation incl. LTC	1000	7400

8.3   Traditional Approach

In the traditional approach the procedures for the calculation of the thermodynamic properties of the working fluid can be divided into the following two main groups:

One-Step Fuel-Oxidation Reaction Mechanism

The one-step fuel-oxidation reaction mechanism is used according to the formulation presented at the beginning of this chapter (very often also the composition variations of rich mixtures due to temperature gradients are neglected). This represents the simplest determination of the composition of the burned gas as basis for the calculation of the thermodynamic properties of the mixture (see 6.2.1.1).

The required CPU-time of the one-step fuel-oxidation reaction mechanisms depending on the chosen numerical procedure can be very convenient. In particular, in case of a premixed mixture with homogeneous air/fuel ratio it is useful to fix "a priori" the composition of the burned gas, instead of a local calculation in each 3D-CFD-cell, and proceed directly with the calculation of the enthalpy as a function of the temperature alone. Under these circumstances the reliability of the results is acceptable as long as the combustion temperatures are lower than 2400 K.

This procedure obviously does not provide information about the reaction speed or, in this case more proper, about the conversion speed of the fresh charge directly into burned gas. Therefore an additional model usually based on a flame propagation calculation is required for the determination of the heat-release.

Reduced Mechanisms of Detailed Reaction Schemes

Starting from the approach of a detailed chemical analysis of all the species present in the working fluid it is useful, in terms of the CPU-time, to find a strategy that permits to substitute the original scheme with a new one described by a limited number of relevant species. This can be attempted by using reduced reaction mechanisms, based mainly on quasi-steady state and partial-equilibrium assumptions so that the influence of many intermediate products can be neglected [55]. A reduced mechanism is still a comprehensive combustion model that theoretically allows a combined calculation of both the evolution of the flame propagation and the related changing in the thermodynamic properties of the working fluid.

This procedure is very complex, the CPU-time is still extremely high and the reliability of the results depends on the assumptions made in setting the reduced mechanisms. As widely reported in the literature (e.g. [55]) such reduced mechanisms are usually devised for certain conditions, i.e. they provide reliable results only for a certain range of temperature, pressure and mixture composition. Since these mechanisms do not have a general validity and combustion processes in internal combustion engines take place under significant thermodynamic gradients, this kind of approach is of low interest in 3D-CFD-simulations for practical applications.

8.4 *QuickSim's* Approach: Few Species for the Description of the Working Fluid

The target of this approach is a convenient and CPU-time-efficient description of the thermodynamic properties of the working fluid. The proposed formulation takes the following points into account:

- Description of the fresh charge and the exhaust gas for any $C_nH_mO_rN_q$ fuel and LHV (Lower heating value)

- Local mixture inhomogeneities

- Dissociation effects and other mixture composition variations at high temperatures (T >1700 K)

- Post-oxidation within the exhaust gas at high temperatures ($T > 1700$ K)

- Post-oxidation between exhaust gas and fresh gas

- Few species (numerical scalars) for the description of the mixture

- Fast calculation of the thermodynamic properties thanks to comprehensive databases or trained neural networks

In Figure 8.3 a schematic of the proposed approach in *QuickSim* is presented.

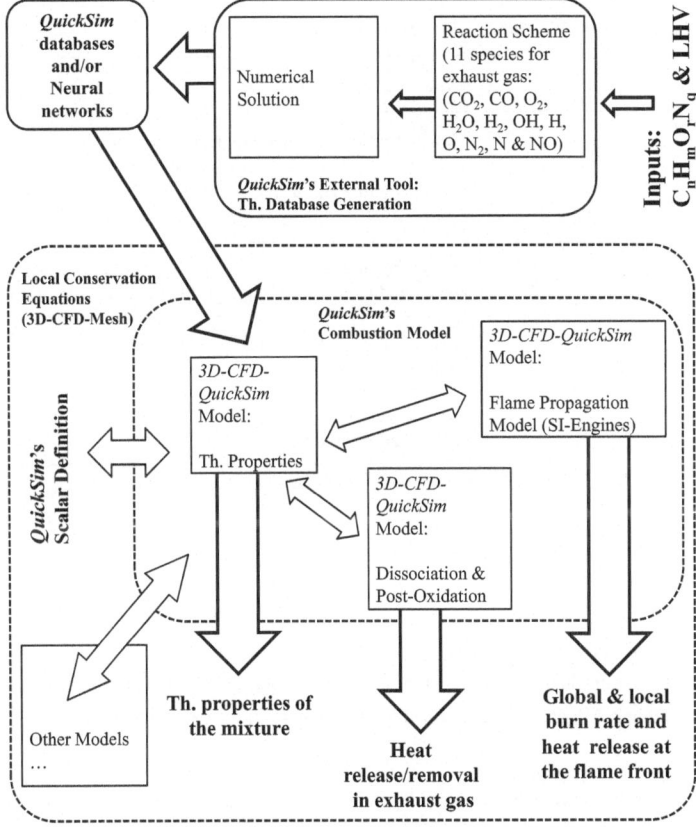

Figure 8.3: *QuickSim's Combustion model: An efficient separation between the composition of the working fluid and the reaction speed (flame propagation speed for SI-engines).*

According to this approach the modeling of the combustion process at the flame front in *QuickSim* is performed by two separated models for the calculation of the thermodynamic properties (with own reaction mechanisms) and the flame propagation, respectively. An additional model implemented within the exhaust gas zone sets the heat release due to post-oxidation (e.g. mixing of exhaust gases with different values of local air/fuel ratios) and the heat exchange (release or removal) due to dissociation effects.

Table 8.2: Scalar definitions in the 3D-CFD-cell *j* using *QuickSim*.

Scalar No.	Variable Name	Description
1	$w_{Air\_U,j}$	Mass fraction of fresh air
2	$w_{F\_U,j}$	Mass fraction of fresh vaporized fuel
3	$w_{EGR\_Air\_U,j}$	Mass fraction of air that has previously produced EGR (burned gas of the previous operating cycle)
4	$w_{EGR\_F\_U,j}$	Mass fraction of vaporized fuel that has previously produced EGR (burned gas of the previous operating cycle)
5	$w_{Air\_B,j}$	Mass fraction of air that has previously produced burned gas
6	$w_{F\_B,j}$	Mass fraction of vaporized fuel that has previously produced burned gas

The description of the mixture composition is based on six scalars (Table 8.2), which have to be considered as the minimalistic choice for describing a reactive working fluid based on an inhomogeneous mixture. In the past, in case of a premixed homogeneous mixture *QuickSim* has worked with only three scalars (fresh gas, EGR and burned gas; all with a fixed air/fuel ratio). Successively the introduction of an injector model has required the definition of four scalars (air, fuel, EGR and burned gas; with a fixed value of air/fuel ratio for the exhaust gases).

Nowadays the standard implementation of six scalars does not require imposing a homogeneous mixture and combustion anymore, i.e. independently on the engine type (MPI, GDI, Diesel, etc.) the approach in describing the working fluid remains the same.

In this approach the scalars are not directly related to a well defined species (chemical compound) anymore but they are trimmed and optimized for identifying and describing the working fluid in its main groups of interest (fresh gas, EGR and burned gas) during all the phases of the engine operating cycle (see Figure 8.4). Consequently the modeling of all engine processes (in particular the combustion model) has to be adapted and also optimized to this formulation, so that, at the end, the 3D-CFD-simulation can gain in efficiency and in reliability.

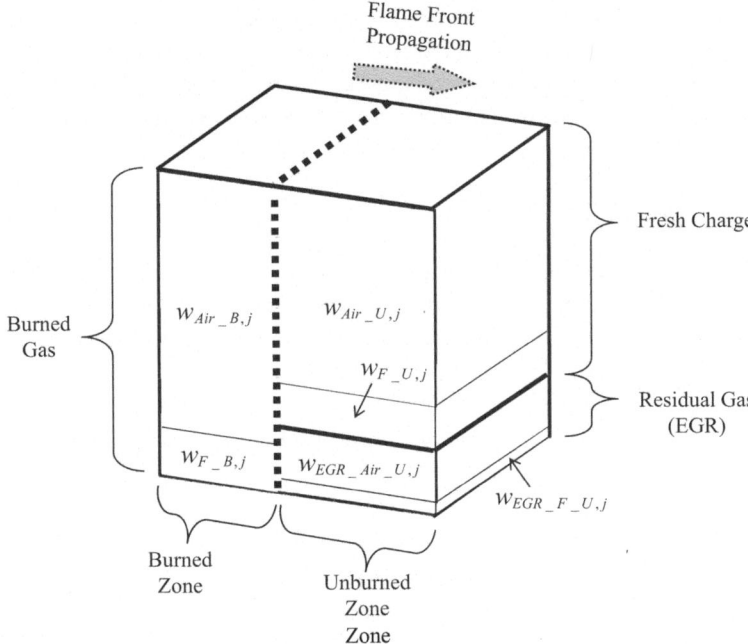

Figure 8.4: *Definition of the composition of the working fluid in the 3D-CFD-cell j of "QuickSim" using six scalars.*

<u>Burned Gas and EGR, two different Exhaust-Gases</u>

The burned gas and EGR, despite the same origin as exhaust gases, are taken separated. This procedure is very helpful during the combustion process, because EGR describes exhaust gas in front of the flame at unburned temperature, whereas burned gas is the newly generated exhaust gas at burned temperature due to combustion. Based on this approach the flame front in a CFD-cell can be easily identified and the local repartition between unburned and burned zone is unequivocally defined. Moreover the mass fraction of EGR is a fundamental variable in the calculation of the laminar flame speed S_L (see Eq. 4.17).

After the end of the combustion (usually at EVO), when the temperatures in the combustion chamber are lower and the mixture composition variations due to temperature gradient can be neglected, a "resetting step" is performed and the burned gas is converted into EGR building the new unburned zone for the next cycle.

The separation of burned gas and EGR permits also the implementation of different approaches for the calculation of the thermodynamic properties that very often allows an optimization of the calculation time. E.g. in case of a well homogenous combustion, after the "resetting step", due also to lower pressure, it may be convenient to use a very simplified and time-efficient database, that takes only the effect of the temperature into account for the calculation of the thermodynamic variables.

Derivated Variables

The first relevant relation is the species conservation equation according to Eq. 6.5. For the numerical actuation of both the combustion and the "resetting" of EGR at the end of the operating cycle the production or sink s_f term for each species involved will be correspondingly modeled.

$$\sum_{i=1}^{6} w_{i,j} = w_{Air\_U,j} + w_{F\_U,j} + w_{EGR\_Air,j} + w_{EGR\_F,j} + w_{Air\_B,j} + w_{F\_B,j} =$$
$$w_{Fresh,j} + w_{EGR,j} + w_{B,j} = w_{U,j} + w_{B,j} = 1 . \tag{8.7}$$

From the local species mass-fractions in each cell, global variables (e.g. for the combustion chamber with N_{Cells}) can be easily calculated. Exemplarily the air mass becomes:

$$m_{Air} = \sum_{j=1}^{N_{Cells}} \rho_j \cdot w_{Air\_U,j} \cdot dV_j . \tag{8.8}$$

Other relevant local variables of interest in internal combustion engines can then be derived from the mass fractions of the starting six species:

$$w_{Fresh,j} = w_{Air\_U,j} + w_{F\_U,j} \quad \text{and} \quad \lambda_{Fresh,j} = \frac{w_{Air\_U,j}/w_{F\_U,j}}{L_{min}} \tag{8.9}$$

$$w_{EGR,j} = w_{EGR\_Air,j} + w_{EGR\_F,j} \quad \text{and} \quad \lambda_{EGR,j} = \frac{w_{EGR\_Air,j}/w_{EGR\_F,j}}{L_{min}} \tag{8.10}$$

$$w_{B,j} = w_{Air\_B,j} + w_{F\_B,j} \quad \text{and} \quad \lambda_{B,j} = \frac{w_{Air\_B,j}/w_{F\_B,j}}{L_{min}} \tag{8.11}$$

$$w_{U,j} = w_{Fresh,j} + w_{EGR,j} \tag{8.12}$$

$$w_{EGR\_U,j} = \frac{w_{EGR,j}}{w_{U,j}} \cdot \qquad (8.13)$$

In particular the mass fraction of EGR related to the unburned zone (i.e. in front of the flame) is, as mentioned before, a variable of primary importance for combustion models.

As discussed at the beginning of this chapter the thermodynamic properties of a combustion gas, first of all depends on its chemical composition, i.e. for any arbitrary fuel $C_nH_mO_rN_q$ as the result of complex mechanisms. For this reason, the definition of lambda values (or air/fuel ratios ϕ) here is determined by two scalars for burned gas and EGR, respectively, and in combination with pressure and temperature introduces a necessary strategy that helps *QuickSim* bridging the information leak over a detailed chemical analysis (see Eq. 8.14). This "dynamic" modeling of the properties of the combustion products is relevant for the final description of the working fluid. The models used for this task are assumed to approach as much as possible the properties of the real gas from the thermodynamic point of view, i.e. the calculated fluid composition that also may include pollutant species (NO_X and HC) cannot be properly used for the determination of exhaust emissions.

The dynamic modeling of combustion products is consequential described by a formulation based on databases or trained neural networks, etc (see Figure 8.3), e.g. for the determination of the thermal enthalpy of the gas in the burned zone of cell j:

$$h_{B,j} = f(\lambda_{B,j}, p_j, T_j) \qquad (8.14)$$

Chapters 8.4.1 and 8.4.2 will explain how these functions can be conveniently established and how this approach can be indistinguishably used for the 3D-CFD-simulation and the real working-process analysis WP [12,33,34,35].

8.4.1   *QuickSim's* Approach: A universally-valid Chemical Reaction Scheme for the Description of Burned Gas

The first step towards the thermodynamic description of the burned gas, as introduced at the beginning of this chapter, is the determination of its chemical composition using a reaction scheme. For this task the atom-numbers n,m,r and q of the fuel $C_nH_mO_rN_q$ and the λ value of the fresh charge have to be known.

Considering in general a chemical reaction that involves the species A, B, C, ... the relationship among educts and products can be formally described as follows [55]:

$$a \cdot A + b \cdot B + c \cdot C + \dots \quad \xrightarrow{\quad k^{(f)} \quad} \quad d \cdot D + e \cdot E + f \cdot F + \dots \qquad (8.15)$$

so that an empirical formulation for the reaction speed of any species concentration, e.g. $[A]$ (mol/m^3), can be introduced:

$$\frac{d[A]}{dt} = -a \cdot k^{(f)} \cdot [A]^a \cdot [B]^b \cdot [C]^c \cdot \dots \qquad (8.16)$$

where $k^{(f)}$ is the reaction speed coefficient of the forward reaction (f). Similarly to the forward reaction also the backward reaction (b) can be introduced:

$$a \cdot A + b \cdot B + c \cdot C + \dots \quad \xleftarrow{\quad k^{(b)} \quad} \quad d \cdot D + e \cdot E + f \cdot F + \dots \qquad (8.17)$$

where $k^{(b)}$ is the reaction speed coefficient of the backward reaction. The variation of the species concentration $[A]$ due to the backward reaction consequently is:

$$\frac{d[A]}{dt} = a \cdot k^{(b)} \cdot [D]^d \cdot [E]^e \cdot [F]^f \cdot \dots \qquad (8.18)$$

Since the forward and backward reactions proceed contemporaneously the resulting variation of the species concentration becomes:

$$\frac{d[A]}{dt} = a \cdot \left(-k^{(f)} \cdot [A]^a \cdot [B]^b \cdot [C]^c \cdot \dots \; + k^{(b)} \cdot [D]^d \cdot [E]^e \cdot [F]^f \cdot \dots \right) \qquad (8.19)$$

The reaction speed coefficients $k^{(f)}$ and $k^{(b)}$ are usually described with the empiric Arrhenius-equation:

$$k = A \cdot T^b \cdot e^{-\frac{E_a}{\Re \cdot T}} \qquad (8.20)$$

where A is a pre-exponential factor and E_a the activation energy.

The reaction speed coefficients introduce the concept of reaction kinetics, i.e. the evolution time of a chemical reaction within a more complex reaction scheme. Depending on the time scale of the investigated phenomena, very often the time available to the chemical reactions is enough for reaching an equilibrium among forward and backwards reactions, i.e. the mixture composition. In spark-ignition engines the chemical equilibrium for thermodynamic purposes (e.g. the description of the working fluid) can be usually assumed. Accuracy reduction may occur mainly in case of a highly stratified mixture in the lean zone that is characterized by a low combustion temperature.

8.4.1.1   Chemical Equilibrium Assumption

Assuming a chemical equilibrium among the species in a given reaction scheme, the rates of the forward and backward reactions are of equal magnitude and therefore the concentrations of the educts and products are constant over a relevant time scale for the process investigation in this task:

$$\frac{d[A]}{dt} \cong 0. \tag{8.21}$$

This assumption drastically simplifies the solution of the reaction scheme and helps introducing equilibrium constants K_c and K_p for any reaction as function of the material properties of the species involved in the reactions:

$$K_c = \frac{k^{(f)}}{k^{(r)}} = \frac{[D]^d \cdot [E]^e \cdot [F]^f \cdot \ldots}{[A]^a \cdot [B]^b \cdot [C]^c \cdot \ldots}. \tag{8.22}$$

From the definition of the partial pressure p_i, e.g. ($i{=}A$) for the species $[A]$:

$$p_A = x_A \cdot p = \frac{n_A}{n} \cdot p = [A] \cdot \Re \cdot T \tag{8.23}$$

it follows:

$$K_p = \frac{p_D^d \cdot p_E^e \cdot p_F^f \cdot \ldots}{p_A^a \cdot p_B^b \cdot p_C^c \cdot \ldots} = K_c \cdot \left(\Re \cdot T \right)^{(d+e+f+\ldots -a-b-c-\ldots)}. \tag{8.24}$$

Similar to the Arrhenius equation the equilibrium constant K_p can be calculated in the following form:

$$K_p = e^{\frac{-\Delta_R \overline{G}^0}{\Re \cdot T}}. \tag{8.25}$$

The variable $\Delta_R \overline{G}^0$ is the change of the molar Gibbs energy ("free energy") of a chemical reaction. It can be determined from the molar Gibbs energies of the individual species:

$$\Delta_R \overline{G}^0 = \underbrace{a \cdot g_A + b \cdot g_B + c \cdot g_c + \ldots}_{\text{Educts}} \quad \underbrace{-d \cdot g_D - e \cdot g_E - f \cdot g_F - \ldots}_{\text{Products}} \tag{8.26}$$

where the molar Gibbs energy of a species i is obtained from its molar specific enthalpy of formation $h'_{f,i}$ (J/kmol) and the specific molar entropy s'_i (J/kmol·K):

$$g_i = h'_{f,i} - T \cdot s'_i. \tag{8.27}$$

In order to avoid the calculation of the Gibbs energy of each species i in practical applications it is convenient to use the following formulation with polynomial coefficients $a_{1\_8,j}$ that directly approximate the equilibrium constant $K_{p,j}$ of an arbitrary chemical reaction j of the reaction scheme [5,12,29,31,32,34]:

$$K_{p,j}(T) = 10^{a_{1,j}\cdot\left(\frac{T}{10^4}\right)^7 + a_{2,j}\cdot\left(\frac{T}{10^4}\right)^6 + a_{3,j}\cdot\left(\frac{T}{10^4}\right)^5 + a_{4,j}\cdot\left(\frac{T}{10^4}\right)^4 + a_{5,j}\cdot\left(\frac{T}{10^4}\right)^3 + a_{6,j}\cdot\left(\frac{T}{10^4}\right)^2 + a_{7,j}\cdot\left(\frac{T}{10^4}\right) + a_{8,j}} \qquad (8.28)$$

The polynomial coefficients can be found e.g. in JANAF-tables [57].

8.4.1.2   The proposed Chemical Reaction Scheme

The chosen reaction scheme based on a partial equilibrium assumption is reported here. This scheme includes seven chemical reactions in which the following 11 species are involved: CO_2, CO, O_2, H_2O, H_2, OH, H, O, N_2, N and NO.

1)
$$CO_2 \Leftrightarrow CO + \frac{1}{2}O_2 \qquad (8.29)$$

2)
$$H_2 + \frac{1}{2}O_2 \Leftrightarrow H_2O \qquad (8.30)$$

3)
$$\frac{1}{2}H_2 + \frac{1}{2}O_2 \Leftrightarrow OH \qquad (8.31)$$

4)
$$\frac{1}{2}H_2 \Leftrightarrow H \qquad (8.32)$$

5)
$$\frac{1}{2}O_2 \Leftrightarrow O \qquad (8.33)$$

6)
$$\frac{1}{2}N_2 \Leftrightarrow N \qquad (8.34)$$

7)
$$\frac{1}{2}O_2 + \frac{1}{2}N_2 \Leftrightarrow NO . \qquad (8.35)$$

Particular relevant during the combustion of internal combustion engines are the dissociation processes of CO_2 and H_2O (reactions 1 and 2). The effect of the dissociation of N_2 on the working-fluid composition (reaction 6) is of minor relevance and very often in the working-process analysis, towards a reduction of the calculation time, it can be neglected [12,34,35].

In a 3D-CFD-approach working with a dedicated database (different for any combination of fuel composition $C_nH_mO_rN_q$ and LHV) the calculation time for the generation of the database is not relevant (approximately two hours on a common PC), therefore the dissociation of N_2 has been considered worth for implementation in the reaction scheme.

From Eq. 8.24 it follows for each reaction:

1)
$$p_{CO_2} = \frac{1}{K_{p,1}} \cdot p_{CO} \cdot \sqrt{p_{O_2}} \tag{8.36}$$

2)
$$p_{H_2O} = K_{p,2} \cdot p_{H_2} \cdot \sqrt{p_{O_2}} \tag{8.37}$$

3)
$$p_{OH} = K_{p,3} \cdot \sqrt{p_{O_2}} \cdot \sqrt{p_{H_2}} \tag{8.38}$$

4)
$$p_H = K_{p,4} \cdot \sqrt{p_{H_2}} \tag{8.39}$$

5)
$$p_O = K_{p,5} \cdot \sqrt{p_{O_2}} \tag{8.40}$$

6)
$$p_N = K_{p,6} \cdot \sqrt{p_{N_2}} \tag{8.41}$$

7)
$$p_{NO} = K_{p,7} \cdot \sqrt{p_{N_2}} \cdot \sqrt{p_{O_2}} \quad . \tag{8.42}$$

In addition, the sum of the partial pressures of all species must be equal to the total pressure;

$$p = p_{CO_2} + p_{CO} + p_{O_2} + p_{H_2O} + p_{H_2} + p_{OH} + p_H + p_O + p_{N_2} + p_N + p_{NO} \quad . \tag{8.43}$$

Furthermore, the number of atoms do not vary over the combustion process, i.e. it is convenient to introduce the following atomic ratios:

$$\frac{N_C}{N_O} = \frac{p_{CO_2} + p_{CO}}{2 \cdot p_{CO_2} + p_{CO} + 2 \cdot p_{O_2} + p_{H_2O} + p_{OH} + p_O + p_{NO}} = const. \tag{8.44}$$

$$\frac{N_O}{N_N} = \frac{2 \cdot p_{CO_2} + p_{CO} + 2 \cdot p_{O_2} + p_{H_2O} + p_{OH} + p_O + p_{NO}}{2 \cdot p_{N_2} + p_N + p_{NO}} = const. \tag{8.45}$$

$$\frac{N_H}{N_O} = \frac{2 \cdot p_{H_2O} + 2 \cdot p_{H_2} + p_{OH} + p_H}{2 \cdot p_{CO_2} + p_{CO} + 2 \cdot p_{O_2} + p_{H_2O} + p_{OH} + p_O + p_{NO}} = const \quad . \tag{8.46}$$

These atomic-ratios can then be described from the compositions of the combustion educts. For an arbitrary fuel $C_nH_mO_rN_q$ in a mixture with a relative air/fuel ratio λ they become:

$$\frac{N_C}{N_O} = \frac{n}{r + 2\lambda \cdot \left(n + \frac{m}{4} - \frac{r}{2} \right)} = const. \tag{8.47}$$

$$\frac{N_O}{N_N} = \frac{r + 2\lambda \cdot \left(n + \dfrac{m}{4} - \dfrac{r}{2}\right)}{q + 3.773 \cdot 2\lambda \cdot \left(n + \dfrac{m}{4} - \dfrac{r}{2}\right)} = const. \tag{8.48}$$

$$\frac{N_H}{N_O} = \frac{m}{r + 2\lambda \cdot \left(n + \dfrac{m}{4} - \dfrac{r}{2}\right)} = const \quad . \tag{8.49}$$

Now 11 equations for 11 species are given and the equation system can be solved. In order to solve this nonlinear equation system, usually (as reported in the literature [29-32]) a Newton procedure with Jacobi matrices is recommended. Since this procedure has significant disadvantages with regard to computing time, an "online" calculation during the 3D-CFD-simulation is not affordable. Therefore, as mentioned before, a comprehensive database or a trained neural network for any fuel over the range of interest of temperature, pressure and lambda has to be created. Recently, a new procedure for the solution of a reduced chemical reaction scheme (9 species) has been developed for the real working-process analysis [34,35]. This procedure allows a computational time reduction up to a factor 50, so that an "online" calculation" of the properties of the working fluid becomes affordable.

8.4.1.3   A "frozen" Composition at low Temperatures

The calculation procedure here discussed provides the concentrations of the species assuming chemical equilibrium in the reaction scheme, i.e. as if an infinite amount of reaction time would be available. In engine processes however, the reaction time is limited while at the same time the chemical reaction rates decrease exponentially as temperature drops. Thus in reality, during the expansion stroke, the real concentrations initially "lag" behind as temperature decreases and under a specific temperature level demonstrate no further change ("frozen state" mixture composition).

The choice of a freeze temperature T_{freeze} as "switch" temperature between equilibrium and frozen composition (for $T \le T_{freeze}$ the reference equilibrium temperature remains T_{freeze}) is a common simplification in this approach. In the literature the reported figures for the "freeze temperature" fluctuate between 1000K and 1900K [30]. The choice of the freeze temperature is in fact driven by the following considerations:

- A general and constant "freeze temperature" must always be a compromise since the freeze temperature is dependent on the engine, on the combustion process and also on the model assumptions.

- Very low freeze temperatures do indeed create a reliably frozen state and thus give rise to better results during the charge exchange period, but also to even greater errors during the working period.

Comprehensive reaction-kinetic analyses at the FKFS have shown that $T_{freeze} = 1600K$ in the working-process analysis and $T_{freeze} = 1700K$ in the 3D-CFD-simulation are a good compromise [12,31,32,34,35].

8.4.1.4   Results: The Chemical Composition of Burned Gas

The solution of the reaction scheme presented in Chapter 8.4.1.2 provides the partial pressures of the 11 species that build the burned gas. Here it is convenient to convert the partial pressure p_i into the corresponding mass fraction w_i (kg/kg):

$$w_i = p_i \cdot \frac{M_i}{M \cdot p} = p_i \cdot \frac{R}{R_i \cdot p} \qquad (8.50)$$

The following diagrams show the composition of the burned gas for different values of temperature, pressure and lambda $\lambda = 1/\phi$. The fuel used is the commercial gasoline ARAL Ultimate 100 (RON=100) : $C_{7.9}H_{14.2}O_{0.12}N_{0.01}$ with LHW = 42.32 MJ/kg.

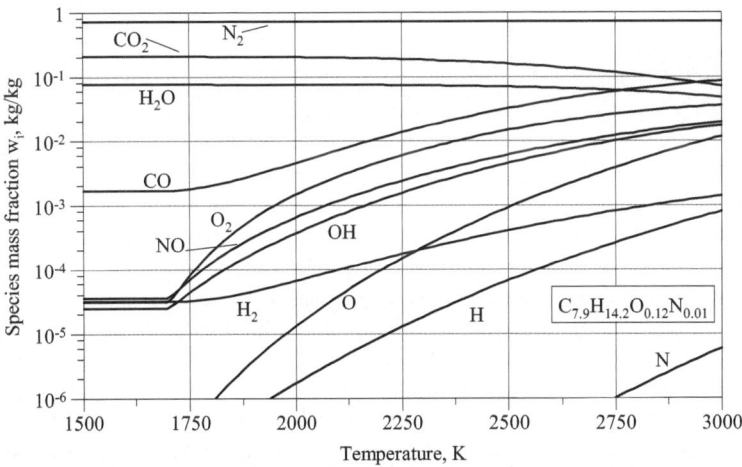

Figure 8.5: *Burned gas composition as function of temperature - $\lambda = 1$ and $p = 1 bar$. Fuel: commercial gasoline with RON 100.*

As expected, Figure 8.5 shows an exponential increasing of the concentration of light species (H_2, OH, H, O, NO and N) with increasing temperature. These "dissociation products" play a relevant role in the typical range of combustion temperatures of an internal combustion engine (T up to ca. 3000K). The dissociation product N, actually, has a limited influence on the mixture composition and often, as reported in other works [12,35], can be excluded from the reaction scheme (elimination of chemical reaction 6 – Eqs. 8.34 and 8.41).

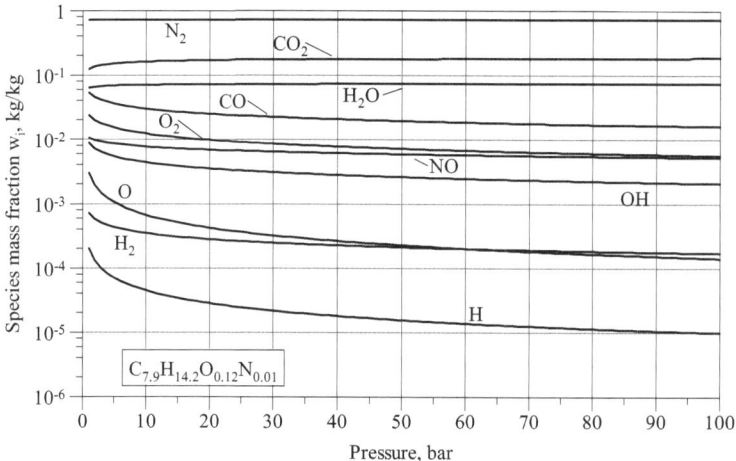

Figure 8.6: *Burned gas composition as function of pressure – T =2700K and λ=1. Fuel: commercial gasoline with RON 100.*

In contrast to temperature, an increasing pressure inhibits dissociation effects (see Figure 8.6). This means there is a less relevant influence of the dissociation during the first half of the combustion duration (up to 50% burned mass) where the pressure in the cylinder at FTDC is very high and there are more relevant dissociation effects during both the second half of the combustion duration and the expansion stroke. Approaching EVO the temperature decreases and when $T \leq T_{freeze}$ the mixture composition becomes "frozen".

The effect of lambda (or air/fuel ratio) on dissociation effects is more complex (see Figure 8.7). Dissociation products based on oxygen (O, NO, and OH) are relevant for lean mixtures but in this case their concentration moderately increases with increasing lambda (air excess). In case of rich mixtures, apart H, the concentrations of the other dissociation products decrease with increasing fuel enrichment and, in general, their influence on the mixture composition is lower than for lean mixtures.

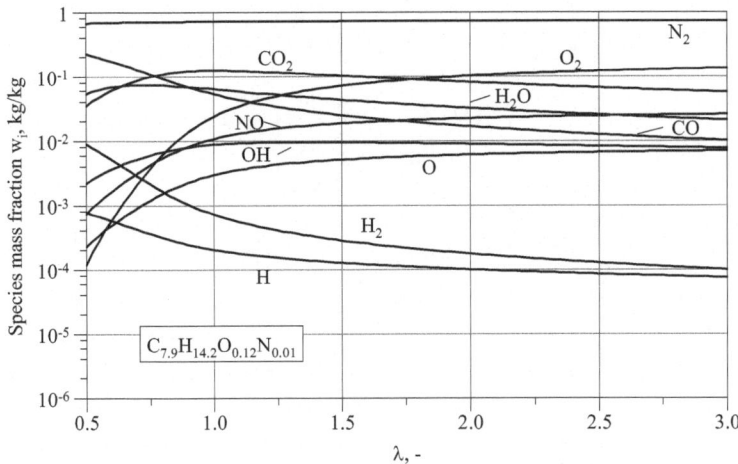

Figure 8.7: *Burned gas composition as function of lambda – T =2700K and p =1 bar. Fuel: commercial gasoline with RON 100.*

Comparison: One-Step Fuel-Oxidation Reaction Mechanism vs. Reaction Scheme

A comparison between the one-step fuel-oxidation reaction mechanism (see Chapter 8.2.1) and the reaction scheme used for *QuickSim* helps understanding the sensitivity of the two approaches on the results. First of all it is interesting to evaluate the dissociation rates (%) of the species CO_2, H_2O and N_2 using the reaction scheme (final composition: w_i) with the following formulation:

$$diss\_rate\_i = \frac{w_{i\_1Step-Ox.} - w_i}{w_{i\_1Step-Ox.}} \cdot 100. \tag{8.51}$$

In Figure 8.8 the dissociation rates as a function of temperature are reported. Dissociation effects are particular relevant for CO_2 followed by H_2O and at the end N_2 with approximately a moderate 10% dissociation rate at $T = 3000K$. These effects become more tangible with the calculation of the real gas constant R (see Figures 8.9-8.11). At low pressure the differences in the calculation of R raise up to 9% at $T = 3000K$ and still are about 3% at high pressure levels and $T = 2700K$. As introduced before, dissociation effects are more sensitive in lean mixtures (e.g. stratified combustion processes), while in a rich mixture these effects decrease with increasing fuel enrichment.

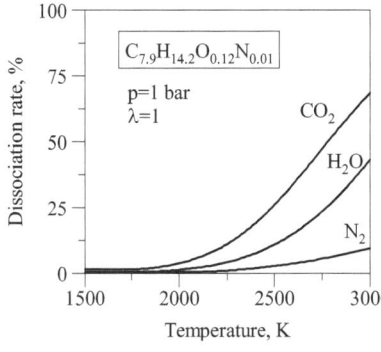

Figure 8.8: *Dissociation rate of CO₂, H₂0 and N₂ in the reaction scheme.*

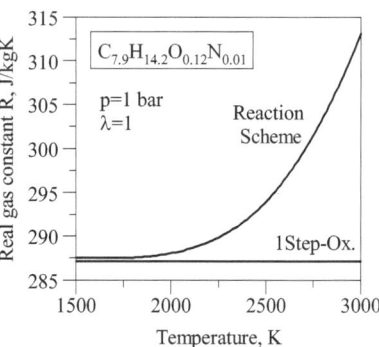

Figure 8.9: *1-Step-oxidation vs. reaction scheme – R=f(T).*

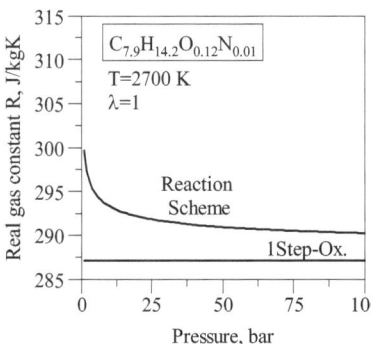

Figure 8.10: *1-Step-oxidation vs. reaction scheme – R=f(p).*

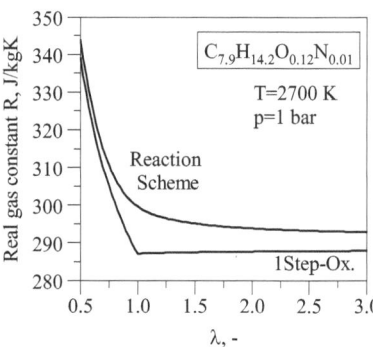

Figure 8.11: *1-Step-oxidation vs. reaction scheme – R=f(λ).*

8.4.2 *QuickSim's* Approach: Conclusive Modeling of the Thermodynamic Properties of Burned Gas

Starting with the composition of the burned gas it is now possible to calculate the thermodynamic properties for the energy conservation equation (see Eq. 6.12). In *QuickSim* a separation of the total enthalpy h_{tc} into the thermal term h and the chemical term h_f (heat of formation) has been conveniently implemented. This procedure handles the flow field as a non-reactive one in combination with an "external" source/sink term s_f that reproduces the heat exchange due to chemical reactions (heat release due to combustion processes at the flame front, post-oxidation and heat-exchange due to dissociation). Accordingly, the energy conservation equation becomes:

$$\frac{\partial(\rho h)}{\partial t} - \frac{\partial p}{\partial t} + \text{div}\left(\rho h \vec{v} + \vec{j}_q\right) + \overline{\overline{P}}: \text{grad}(\vec{v}) - \text{div}(p\vec{v}) = s_f \qquad (8.52)$$

where the source term s_f is:

$$s_f = \frac{D(\rho h_f)}{Dt} = \frac{\partial(\rho h_f)}{\partial t} + \text{div}\left(\rho h_f \vec{v}\right). \qquad (8.53)$$

The thermal enthalpy h and the heat of formation h_f can be easily calculated using polynomial forms with coefficients from JANAF-tables [57] (see Eqs. 6.16-6.18).

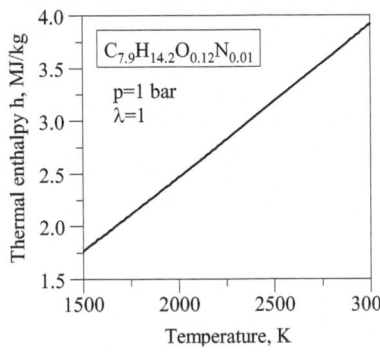

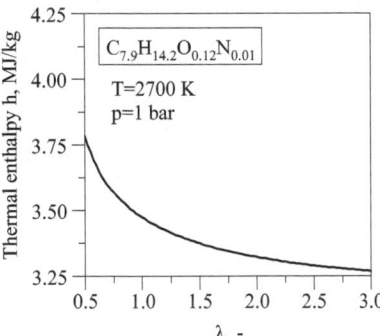

Figure 8.12: *Thermal enthalpy h as function of temperature.*

Figure 8.13: *Thermal enthalpy h as function of lambda.*

In Figure 8.12 and Figure 8.13 the thermal enthalpy h as a function of temperature and lambda is shown. The influence of temperature and lambda on the enthalpy h is remarkable while the influence of pressure is often negligible. Here the differences between the one-step fuel-oxidation reaction mechanism and the reaction scheme are usually very small (<1%).

As introduced before in order to calculate the heat exchange terms related to chemical reactions in generally, it is now mandatory to determine the heat of formation of exhaust gas h_f and fresh gas $h_{f,Fresh}$.

8.4.2.1   Heat Release at the Flame Front and Post-Oxidation of Exhaust Gas with Fresh Gas

The local heat release $\Delta Q_{B,Ox}$ at the flame front ("primary" oxidation of a discrete local amount of fresh gas Δm_{Fresh} during the combustion process) or the post-oxidation $\Delta Q_{B,Ox,Post}$ of exhaust gas $\Delta m_{B,0}$ with fresh gas Δm_{Fresh} ("secondary" oxidation that produces a final exhaust gas with properties "1" – see Eq. 8.55) are described in *QuickSim* as follows:

$$\Delta Q_{B,Ox} = \Delta m_{Fresh} \cdot \left(h_{f,Fresh} - h_f\right) = \Delta m_{Fresh} \cdot \Delta h_{HR} \qquad (8.54)$$

$$\Delta Q_{B,Ox,Post} = \Delta m_{Fresh} \cdot h_{f,Fresh} + \Delta m_{B,0} \cdot h_{f,0} - \left(\Delta m_{Fresh} + \Delta m_{B,0}\right) \cdot h_{f,1} \qquad (8.55)$$

where the heat of formation of the burned gases $h_f, h_{f,0}$ and $h_{f,1}$ are calculated using Eq. 6.18 with a corresponding local fluid composition and the tabled values of the species heat of formation (see Table 8.3 [57]). Similarly the heat of formation of the fresh gas $h_{f,Fresh}$ becomes:

$$h_{f,Fresh} = h_{f,F} \cdot w_F + h_{f,O_2} \cdot w_{O_2} + h_{f,N_2} \cdot w_{N_2} \qquad (8.56)$$

Table 8.3: Heat of formation $h_{f,j}$ of species j (MJ/kg).

CO	H$_2$O	OH	H	O	CO$_2$	O$_2$	H$_2$	N	N$_2$	NO
-3.9475	-13.4339	2.31176	218	15.575	-8.94318	0	0	33.7628	0	3.00966

In *QuickSim* the calculation of the fuel's heat of formation $h_{f,F}$ is based on the lower heating value h_{LHV} as input and is not strictly related to the heat of formations of the species building the implemented fuel: $C_nH_mO_rN_q$. This means that the parameters n,m,r and q can be conveniently chosen as an approximation of a more complex real composition of the fuel and the heat of formation $h_{f,F}$ can be easily calibrated to the effective h_{LHV}. From the definition of the lower heating value h_{LHV} (heat release with complete fuel oxidation at $\lambda=1$ and "frozen" composition):

$$\Delta Q_{B,Ox,Max} = \Delta m_F \cdot h_{LHV} \qquad (8.57)$$

Using Eqs. 8.54 and 8.56 and the relation of a stoichiometric fuel oxidation (see Eq. 8.1) the fuel heat of formation $h_{f,F}$ becomes:

$$h_{f,F} = \frac{h_{LHV} + \left(n \cdot M_{CO_2} \cdot h_{f,CO_2} + \frac{m}{2} \cdot M_{H_2O} \cdot h_{f,H_2O} \right)}{1 \cdot M_F}. \tag{8.58}$$

In Figures 8.14 – 8.17 the heat of formation $h_{f,F}$ of the exhaust gas as a function of temperature, pressure and lambda is reported. As expected, $h_{f,F}$ rises with increasing the effect of dissociation. This effect reduces the heat-release because a part of the energy is still stored in species with high chemical potentials. The comparison between the assumed reaction scheme and a simple one-step fuel-oxidation mechanism shows remarkably differences, which would irremediably influence the heat-release evaluations.

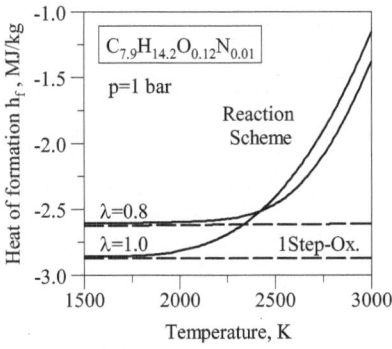

Figure 8.14: *Heat of formation h_f (HOF) as function of temperature (p =1bar).*

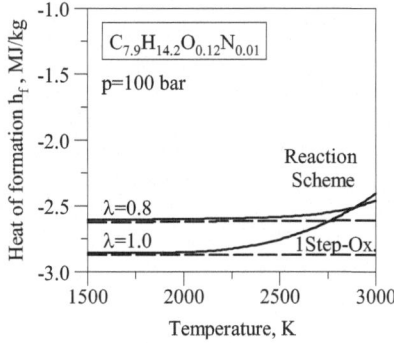

Figure 8.15: *Heat of formation h_f (HOF) as function of temperature (p =100bar).*

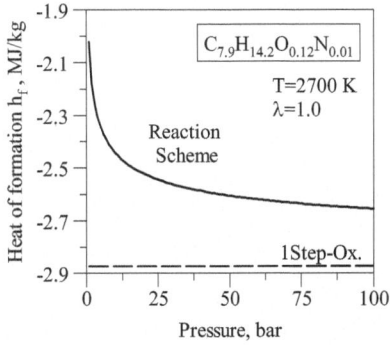

Figure 8.16: *Heat of formation h_f (HOF) as function of pressure.*

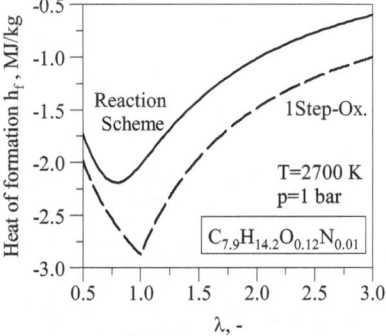

Figure 8.17: *Heat of formation h_f (HOF) as function of lambda.*

8.4.2.2   Heat Exchange due to Dissociation Effects and Post-Oxidation within Exhaust Gas

According to Eq. 8.53 the heat exchange in the burned zone occurs when, due to temperature, pressure and lambda changes, chemical reactions take place. These changes of the chemical composition of the exhaust gas are caused by the fluid's compression and expansion during the working cycle, mixing processes between zones with different air/fuel ratio, etc. Similarly to the procedure in Chapter 8.4.2.1 the variations of the heat of formation h_f (in particular the gradients: dh_f/dT, dh_f/dp and $dh_f/d\lambda$) within the burned zone are used for a local description of the heat-exchange amount.

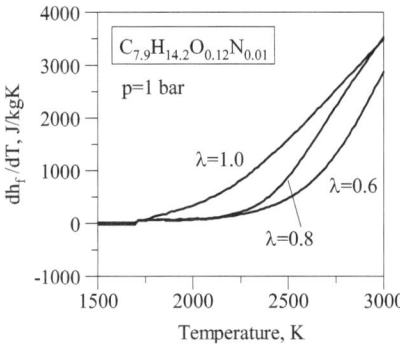

Figure 8.18: Gradient of HOF: dh_f/dT as function of temperature (p =1bar - λ rich).

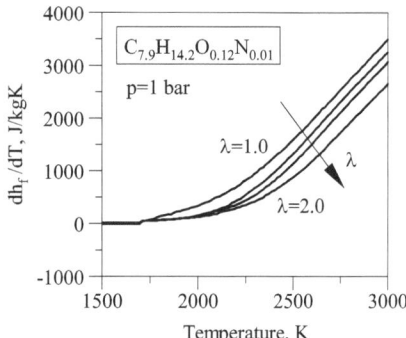

Figure 8.19: Gradient of HOF: dh_f/dT as function of temperature (p =1bar - λ lean).

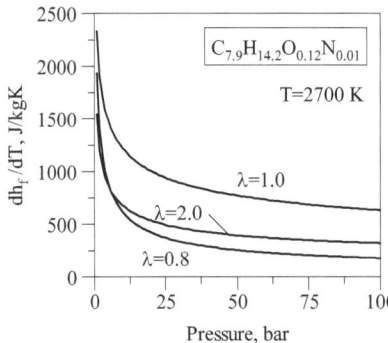

Figure 8.20: Gradient of HOF: dh_f/dT as function of pressure (T =2700K).

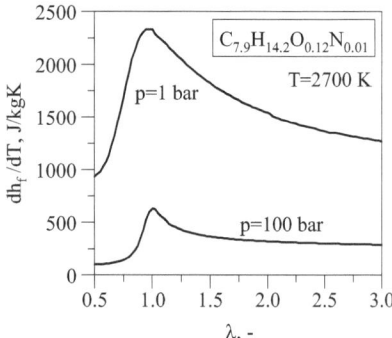

Figure 8.21: Gradient of HOF: dh_f/dT as function of lambda (T =2700K).

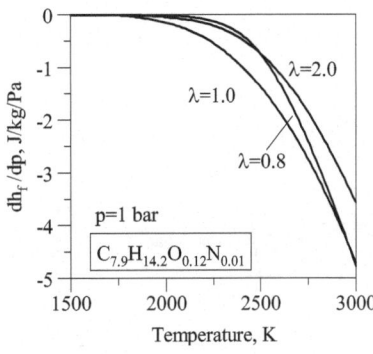

Figure 8.22: *Gradient of HOF: dh_f/dp as function of temperature ($p = 1 bar$).*

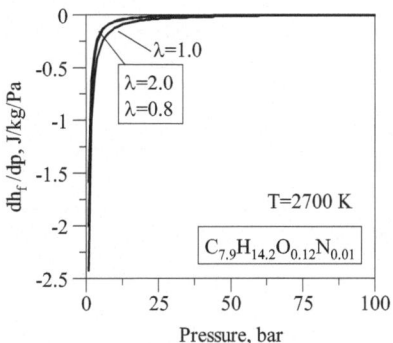

Figure 8.23: *Gradient of HOF: dh_f/dp as function of pressure ($T = 2700K$).*

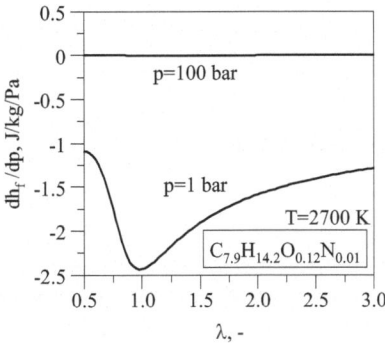

Figure 8.24: *Gradient of HOF: dh_f/dp as function of lambda ($T = 2700K$).*

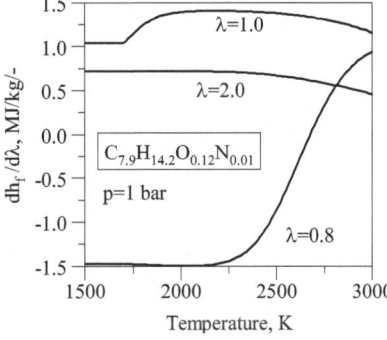

Figure 8.25: *Gradient of HOF: $dh_f/d\lambda$ as function of temperature ($p = 1bar$).*

In Figures 8.18 – 8.27 the gradients of the heat of formation of the burned gas are reported. A positive gradient means the trend of the exhaust gas to convert thermal energy (h) into chemical energy. At low temperatures where the composition is frozen, the gradients dh_f/dT and dh_f/dp are zero. These gradients rise exponentially with increasing the temperature and the influence is more decisive when $\lambda \cong 1$ and the pressure is low. In particular the pressure hinders dissociation effects at any lambda value, i.e. the contribution of the gradients dh_f/dT and dh_f/dp to the heat-exchange drastically decreases as soon as the pressure rises. The contribute of $dh_f/d\lambda$ (mixing processes) to the heat-exchange is of higher complexity. The oxidation of a "rich" burned gas is accompanied by heat release as long as the temperature is not so high that dissociation "absorbs" energy (see Figure 8.25). An increasing pressure has low influence on $dh_f/d\lambda$ for "lean" burned gases but helps the post-oxidation of a "rich" burned gas because dissociation effects slow down.

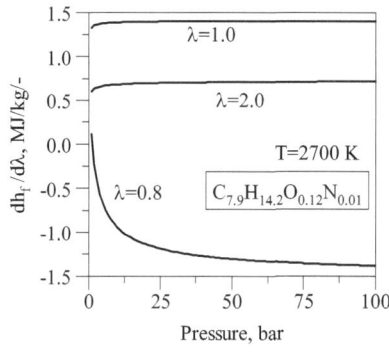
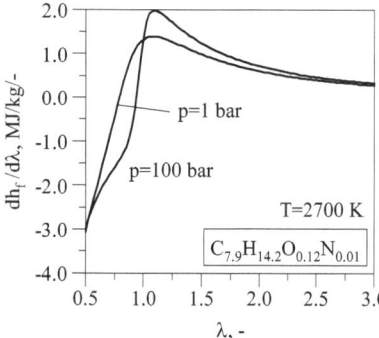

Figure 8.26: *Gradient of HOF: $dh_f/d\lambda$ as function of pressure (T =2700K).*   **Figure 8.27:** *Gradient of HOF: $dh_f/d\lambda$ as function of lambda (T =2700K).*

Mixing Process of Exhaust Gases with and without Heat Exchange

The separation of the thermal and chemical enthalpy to different 3D-CFD-models (see Figure 8.3) for the calculation of the thermodynamic properties of the working fluid and the heat release (combustion at the flame front, post-oxidation and dissociation) allows a selective activation of these models depending on the thermodynamic conditions at which the processes take place. E.g., during the expansion or exhaust stroke a local mixing process at low temperature ($T <1000$ K) between two zones of burned gases ($\Delta m_{B,1}$ and $\Delta m_{B,2}$) with different temperatures and compositions ($w_{Air\_B,1}$, $w_{F\_B,1}$, $w_{Air\_B,2}$ and $w_{F\_B,2}$) leads numerically, due to the conservation equations, to a final exhaust gas $\Delta m_{B,3}$, that, among other things, has the resulting composition:

$$\lambda_{B,1} = \frac{w_{Air\_B,1}/w_{F\_B,1}}{L_{min}} \quad \& \quad \lambda_{B,2} = \frac{w_{Air\_B,2}/w_{F\_B,2}}{L_{min}} \quad \Rightarrow$$

$$\Rightarrow \quad \lambda_{B,3} = \frac{\dfrac{w_{Air\_B,1} \cdot \Delta m_{B,1} + w_{Air\_B,2} \cdot \Delta m_{B,2}}{w_{F\_B,1} \cdot \Delta m_{B,1} + w_{F\_B,2} \cdot \Delta m_{B,2}}}{L_{Min}}. \tag{8.59}$$

The error caused by assuming this final composition as a new chemical equilibrium instead of the sum of two frozen compositions affects moderately the thermal properties of the burned gas but it would be inacceptable if this mixing process generates a heat source, what would be definitely unrealistic. For this reason the code can decide whether a process, due to numerical chemical reactions, is plausibly related to heat-exchange or not.

8.4.2.3   Combustion Conversion Efficiency

According to Eqs. 8.54 and 8.57 a local combustion conversion efficiency η_{HR} can be introduced [12,34]:

$$\eta_{HR} = \frac{\Delta Q_{B,Ox}}{\Delta Q_{B,Ox,Max}} = \frac{\Delta Q_{B,Ox}}{\Delta m_F \cdot h_{LHV}}: \tag{8.60}$$

This efficiency describes the ratio of the oxidation heat release of a given local mixture at a final combustion temperature T to the maximal heat release for a full fuel oxidation, i.e. starting from a local fuel mass Δm_F the efficiency due to *incomplete combustion or incomplete oxidation* can be calculated (see Figures 8.28 and 8.29).

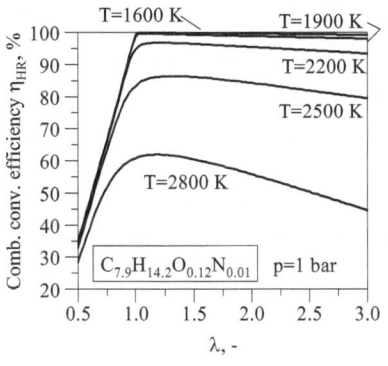

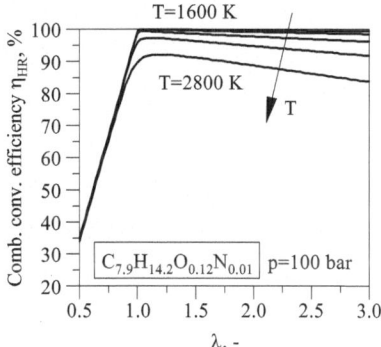

Figure 8.28: *Combustion conversion efficiency η_{HR} (p =1 bar).*

Figure 8.29: *Combustion conversion efficiency η_{HR} (p =100 bar).*

For rich mixtures these figures show a remarkable effective reduction of η_{HR} that can be clearly explained with oxygen deficit. At low temperatures T (frozen state) for $\lambda \geq 1$, a full oxidation is possible so that theoretically $\eta_{HR} = 1$ (an *imperfect combustion* is not taken into account - see 4.4.3). At higher temperatures dissociation effects reduce the heat release, because a part of the chemical energy remains stored in molecules with high chemical potential. This is a fictive reduction of η_{HR}, because as soon as temperature drops the stored "dissociation-energy" can be released. That means that during the working cycle in IC-engines dissociation acts as an energy "capacitor", that is, depending of temperature and pressure, more or less charged during combustion and discharged during the expansion stroke.

9

3D-CFD-Modeling of the Combustion for SI-Engines

In the previous chapter it has been explained that 3D-CFD combustion models are usually heat-release models. These models first calculate the local burn rate and consequently using the thermo-chemical properties of the working fluid the related local heat-release.

Heat-release models in *QuickSim* have been developed for both spark-ignition (based on a burn rate calculation as a function of the flame front propagation within a turbulent partially-premixed mixture) and compression ignition engines (based on both self-ignition mechanisms and diffusive flame propagations). In this chapter the approach in the modeling of the flame propagation within the combustion chamber during the combustion of SI-engines is presented.

9.1 Introduction

Starting from an initial flame kernel generated during the ignition process at the spark plug, the flame propagates up to a certain radius under laminar conditions, then the flame reaches a characteristic dimension at which the interactions with the turbulent eddies support the local oxidation process and the flame speed remarkably increases (more details in [5,55,58,59,61]).

Following the schematic of a turbulent flame propagation in a duct with a section area $A_{f,T}$ (see Figure 9.1) the flame can be modeled as a thin oxidation-sheet wrinkled by turbulent eddies where the behavior of a laminar flame structure remains locally unchanged (flamelet approach). The increasing of the flame speed S_T under turbulent conditions is then explained with the increasing of the effective surface of the wrinkled flame $A_{f,L} > A_{f,T}$ (ensemble of local laminar flames with different flow conditions like: stretch, wrinkling, velocities, etc.). Based on this approach the burn rate rises due to the increasing of the oxidation region.

Following the mass balance the relation between the laminar S_L and the turbulent flame speed S_T can be defined as follows:

$$\rho_U \cdot S_T \cdot A_{f,T} = \rho_U \cdot S_L \cdot A_{f,L} \quad \Rightarrow \quad S_T = S_L \cdot \frac{A_{f,L}}{A_{f,T}} \tag{9.1}$$

where the laminar flame speed $S_L(\lambda_{Fresh,j}, p_j, T_{U,j}, x_{EGR\_U,j})$ in a cell j of the 3D-CFD-mesh is formulated, e.g. with the relations reported in Chapter 4.4.3.2. Figures 9.2, 9.3 and 9.4 show the profiles of S_L as function of lambda, pressure and temperature for a commercial gasoline fuel. As well known the maximal laminar flame speed S_L is reached with a rich mixture at $\lambda_{Fresh} \cong 0.9$.

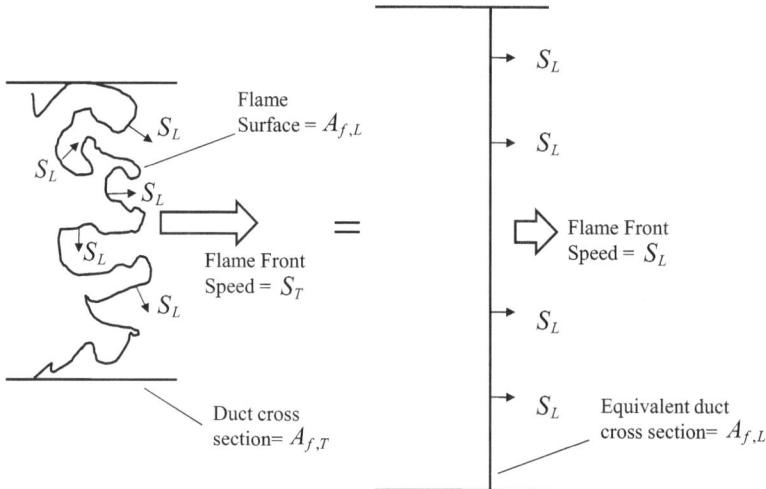

Figure 9.1: *Schematic of turbulent flame propagation (flamelet approach of a wrinkled flame).*

Several models for the description of the wrinkling factor K have been proposed. Despite a lot controversy most of these models recognize the following trend:

$$K = \frac{S_T}{S_L} \sim \frac{u'}{S_L} \tag{9.2}$$

i.e. the turbulence intensity u' is the most sensitive variable for the calculation of the wrinkling factor K and consequentially the turbulent flame speed S_T. Based on this modeling the turbulent flame speed S_T rises with increasing the turbulence intensity u' (see Figure 9.5). In reality it has to be noticed that a continued increasing of u' increases S_T up to a maximum followed by a reduction and, at the end, a flame extinction. An explanation of this behavior can be found considering the local flame strain caused by the turbulence. The Turbulence is responsible for an increasing convection, i.e. among other things, the supply of reactants towards the flame positively increases [55]. On the other hand high convection rates lead to steeper gradients of the flame surface that increase local diffusive losses. When the finite rate of the chemical kinetics is unable to generate products as fast as reactants are delivered and products, including enthalpy are removed by diffusion, the local flame temperature and consequently the reaction rate remarkably decrease. If the flame temperature is too low the whole combustion process becomes unstable and locally the flame suddenly extinguishes.

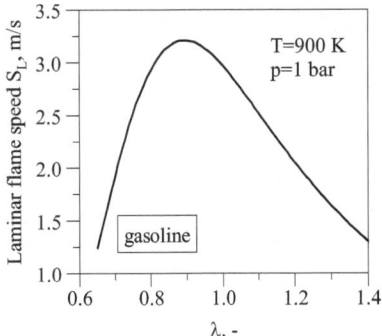

Figure 9.2: *Laminar flame speed S_L as function of lambda (fuel: gasoline).*

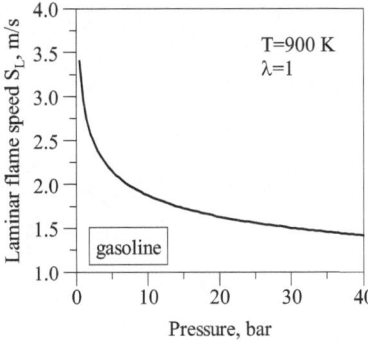

Figure 9.3: *Laminar flame speed S_L as function of pressure (fuel: gasoline).*

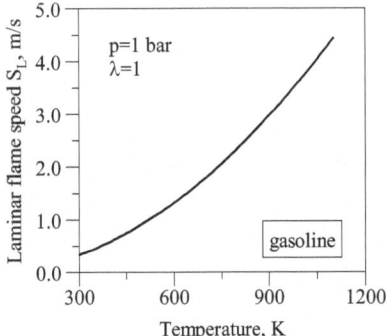

Figure 9.4: *Laminar flame speed S_L as function of temperature (fuel: gasoline).*

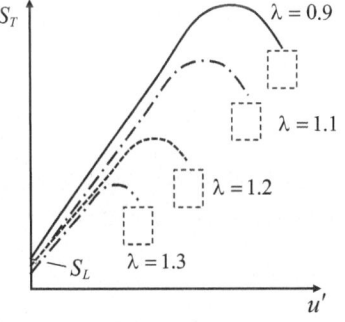

Figure 9.5: *Dependence of turbulent flame speed S_T on the turbulence intensity.*

As reported in Figure 9.5 lean mixtures are particularly sensible to flame extinction. This is due to a combination of several effects like e.g. lower laminar flame speed S_L and lower mixture-specific heat-release density that make the flame propagation more sensitive to local turbulent convection.

9.2 Flame Propagation Modeling (Weller Model)

As introduced before, a developed turbulent flame within a turbulent premixed mixture in SI-engines is a thin reaction sheet (flame front), wrinkled and convoluted by the turbulence of the unburned mixture. Following the one-step flame propagation model proposed by Weller [69] it is convenient to introduce a progression variable c that permits to identify the position of the flame within the 3D-CFD-mesh. Per definition the progress variable c has to assume the value 0 in the unburned zone and 1 in the burned zone (see Figure 9.6).

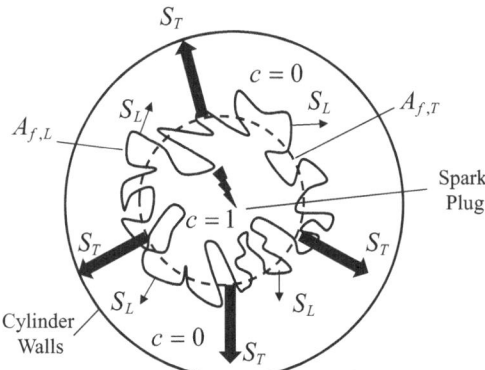

Figure 9.6: *The progress variable c and the flame propagation.*

Several formulations for c have been proposed in the literature [55,59,61], but using the scalar definition of *QuickSim* the progress variable c in the 3D-CFD-cell j can be easily and directly represented by the burned mass-fraction variable $w_{B,j}$ (see Chapter 8.4):

$$c_j = w_{B,j} \qquad (9.3)$$

The species conservation law (see Eqs. 6.38 and 6.41) as a transport equation can then be properly used for describing the flame propagation:

$$\frac{\partial(\overline{\rho}\widetilde{w}_B)}{\partial t} + div(\overline{\rho}\widetilde{w}_B\widetilde{v}) - div(\overline{\rho}D_{B,T}^M \cdot grad\,\widetilde{w}_B) = \overline{M_B\omega_B} = \widetilde{r}_B \qquad (9.4)$$

where $\widetilde{r}_B$ is the local reaction rate of burned gas. During the flame propagation the local combustion rate in the cell j is then defined as follows:

$$\widetilde{r}_{B,j} \cdot V_j = \rho_{U,j} \cdot A_{f,T,j} \cdot S_{T,j} \,. \qquad (9.5)$$

Here it becomes evident that the local turbulent flame surface $A_{f,T,j}$ within a cell is not available numerically. Though, the Weller model describes pragmatically the surface in the cell as a function of the gradient of the progress variable:

$$\widetilde{r}_{B,j} = \rho_j \cdot \frac{dw_{B,j}}{dt} = \rho_{U,j} \cdot S_{T,j} \cdot \left|grad\,w_{B,j}\right| \,. \qquad (9.6)$$

Many formulations for S_T have been proposed during the years. In this work a semi-empirical formulation of the wrinkling factor K proposed by Herweg and Maly [70] has been implemented:

$$K = \frac{S_T}{S_L} = \underbrace{I_0 + I_0^{1/2}}_{Term1} \cdot \underbrace{\left(\frac{\overline{v}_T}{\overline{v}_T + S_L}\right)^{1/2}}_{Term2} \cdot \underbrace{\left(1 - \exp\left(-\frac{r_K}{l_l}\right)\right)^{1/2}}_{Term3} \cdot$$
$$\cdot \underbrace{\left[1 - \exp\left(-\frac{\overline{v}_T + S_L}{l_l} \cdot (t - t_{IP})\right)\right]^{1/2}}_{Term4} \cdot \underbrace{\left(\frac{u'}{S_L}\right)^{5/6}}_{Term5} \qquad (9.7)$$

where I_0 (1st term of K) is the effect of strain on the laminar burning speed at small radiuses (usually $I_0 = 1$), l_l the integral length scale of the flow field (see Chapter 6.2.1.3), r_K the flame kernel radius and $(t - t_{IP})$ the time after the spark ignition IP.

The velocity $\overline{v}_T$ is described as follows:

$$\overline{v}_T = \left(\left|\overline{v}\right|^2 + u'^2\right)^{1/2} \qquad (9.8)$$

It represents the relevant velocity for the wrinkling factor, derived from the sum of the kinetic energy of the main fluid, the velocity $\overline{v}$ and the turbulence k. Assuming an isotropic turbulent field, the turbulence intensity u' is given by:

$$u' = \sqrt{\frac{2 \cdot k}{3}} \qquad (9.9)$$

The other terms of the formulation of K take the following effects into account:

- 2nd term (effective turbulent factor): This factor determines the behavior in the critical region $0 < u'/S_L < 1$ where $\bar{v}$ plays the controlling role.

- 3rd and 4th term (size and time dependent integral scale): These terms taking into account the turbulent length and time scales, respectively, influence the flame kernel development from the very beginning. The turbulence effects become more and more important with increasing size and life time of the flame kernel until the full spectrum of turbulence affects the flame propagation (typical orders of magnitude: $l_l \approx 3\text{mm}$ and $(t - t_{IP}) \approx$ 1ms).

- 5th term (fully developed turbulent combustion): This term describes the behavior of a fully developed, freely expanding flame in an isotropic turbulent flow field.

9.3   *QuickSim*'s Approach: Implementation Improvement

Based on the flame propagation model approach it is then possible to calculate the variations of all scalars that describe the properties of the working fluid in *QuickSim*:

$$\frac{dw_{B,j}}{d\varphi} = -\frac{dw_{U,j}}{d\varphi} \tag{9.10}$$

$$\frac{dw_{B,j}}{d\varphi} = -\left(\frac{dw_{Fresh,j}}{d\varphi} + \frac{dw_{EGR,j}}{d\varphi}\right) \tag{9.11}$$

$$\frac{dw_{Air\_B,j}}{d\varphi} = -\left(\frac{dw_{Air\_U,j}}{d\varphi} + \frac{dw_{EGR\_Air\_U,j}}{d\varphi}\right) \tag{9.12}$$

$$\frac{dw_{F\_B,j}}{d\varphi} = -\left(\frac{dw_{F\_U,j}}{d\varphi} + \frac{dw_{F\_Air\_U,j}}{d\varphi}\right) \tag{9.13}$$

These relations link the mass balance among the species while the energy balance refers to the heat-release formulation introduced in Chapter 8.

9.3.1   Numerical Implementation of the Flame Propagation Model

The numerical implementation of the Weller model implies the application of a mathematical model within discrete elements (cells) which have very often discretization lengths many times bigger than the characteristic lengths of the investigated phenomenon. In order to ensure a low mesh-dependent implementation of the model, first of all, investigations on the sensitivity of the input variables on the results have to be performed (a model is not reliable when its input variables are not reliable). As it will be discussed in this and also in the next chapter not rarely local variables are affected by the local mesh structure, dimension, etc. For this reason very often it is more convenient and numerically more "stable" to take into account global values (averaged over the whole combustion chamber or sometimes parts of it), i.e. a loss of local information as an input in the model is much better than meaningless final results.

In this chapter two meshes of different engines are used for presenting few results (see Figures 9.7 and 9.8):

- One cylinder of a mass-production passenger car engine with two valves ($V = 575$ cm^3, bore 95 mm, asymmetric bowl shape with big squish area, coarse discretization with N_{Cells} =15,000 cells for the combustion chamber alone).

- The cylinder of a mass-production bike engine with four valves ($V = 650$ cm^3, bore 100 mm, symmetric pent-roof chamber with squish areas, coarse discretization with N_{Cells} =25,000 cells for the combustion chamber alone).

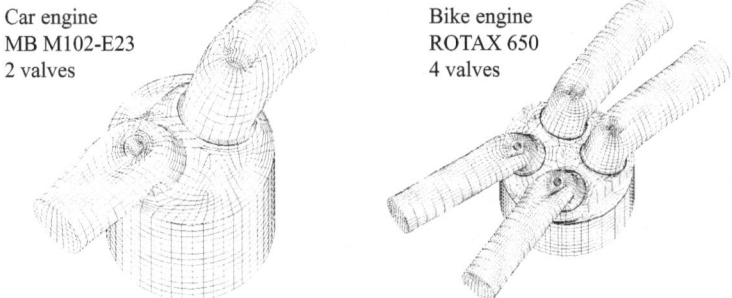

Car engine
MB M102-E23
2 valves

Bike engine
ROTAX 650
4 valves

Figure 9.7: *Calculation mesh of a passenger car engine (MB M102 E23 – 2 valves).*   **Figure 9.8:** *Calculation mesh of a bike engine (Rotax 650 – 4 valves).*

Investigations on different approaches in the implementation of the chosen combustion model require first of all analyses of the processes under the same "thermodynamic conditions", so that the influence of each input variable can be better isolated from other factors. This has been achieved by imposing burn profiles to the 3D-CFD-simulations (see e.g. Figure 9.9), which have been determined by the real working-process analysis calibrated with measured pressure profiles.

With the setting of target burn rate profiles approximated by Wiebe formulations (see Chapter 4.4.3) consequently target heat releases are imposed to the 3D-CFD-simulations, i.e. as assumed in this approach for reliable investigations on the model implementation the operating cycles between experiments and 3D-CFD-simulations are thermodynamically equivalent.

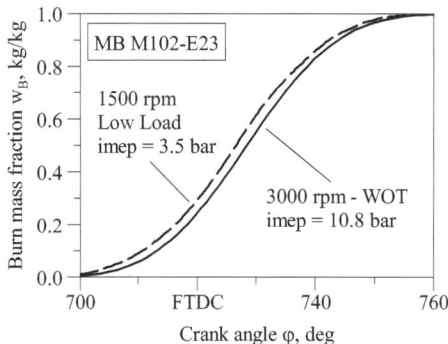

Figure 9.9: *Burn mass fraction profile at different operating conditions (MB M102 E23).*

Target

A target heat release $(dQ_B/d\varphi)_{Exp.}$ is performed in the 3D-CFD-code *QuickSim* using its evaluation tool for the determination first of $(dw_B/d\varphi)_{CFD}$ (see Eq. 9.14) and then $(dQ_B/d\varphi)_{CFD}$ in a "closed loop" for a comparison with experimental data as inputs (see Figure 9.10).

$$\left(\frac{dw_B}{d\varphi}\right)_{CFD} = \frac{1}{6n} \cdot \frac{\sum_{j=1}^{N_{Cells}} \left|\tilde{r}_{B,j}\right| \cdot V_j}{\sum_{j=1}^{N_{Cells}} m_j} \tag{9.14}$$

$$c_{HR}(\varphi) = \left(\frac{dQ_B}{d\varphi}\right)_{Exp.} \Bigg/ \left(\frac{dQ_B}{d\varphi}\right)_{CFD} \tag{9.15}$$

$$\tilde{r}_{B,j,corr} = c_{HR} \cdot \rho_{U,j} \cdot S_{T,j} \cdot \left|grad\ w_{B,j}\right| . \tag{9.16}$$

At any time-step the target heat release is automatically compared with the simulated one and if they differ a global heat-release correction factor $c_{HR}(\varphi)$ is calculated and provided to the 3D-CFD-simulation as a feedback (see Eq. 9.15). This factor is then used iteratively for adjusting the calculation of the local reaction rate of the burned gas (see Eq. 9.16), so that at the end the target profile is matched with high accuracy.

Investigations on the Implementation Phase

Similarly to the heat-transfer investigations (see Chapter 10) in the development phase of a combustion model (or in this case the procedure for an improved implementation of an existing one towards less mesh-dependence) the variation of $c_{HR}(\varphi)$ during the combustion should become very small, so that $c_{HR}(\varphi)$ can be implemented as a constant in the combustion model formulation.

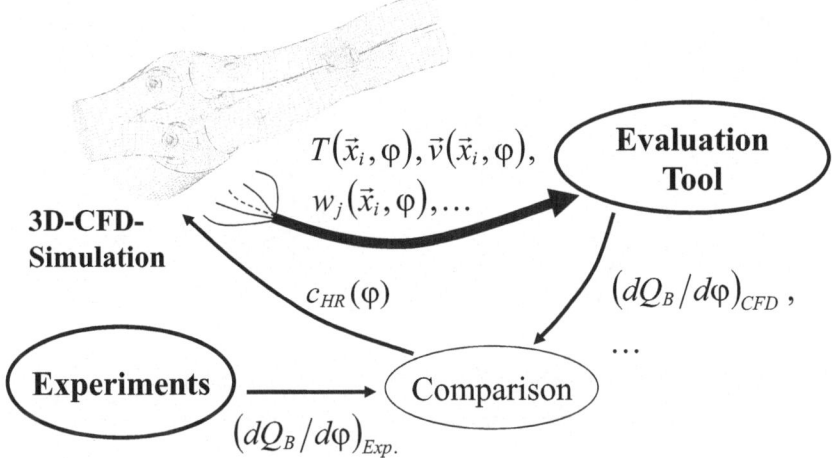

Figure 9.10: *The "internal coupling" between a 3D-CFD-code and experimental data for burn rate control during the implementation phase of the combustion model.*

Starting with the investigation of input variable reliability on the model results, it becomes clear that in the 3D-CFD-simulation few of these variables are per definition not available at all. The problem, as discussed below, resides in the numerical discretization of the flame front.

9.3.2   Numerical Inconsistencies at the Flame Front

When a combustion process takes place in the cylinder, i.e. a flame front is present or self-ignition processes occur, high gradients of temperature are generated due to the local heat release. The oxidation energy rises the temperature of the already burned mixture up to a temperature difference of ca. 2000 K to the unburned zone; while heat transfer through the border of the burned zone slowly warms the unburned zone. A cell j of the 3D-CFD-mesh involved in the oxidation process at the flame front (SI-engines), i.e. where the burned mass fraction $0 < w_{B,j} < 1$, shows a cell temperature $T_P(\vec{x}, \varphi)$ (per definition in the central node of the cell P – see Figure 9.11) with a value in-between the unburned and burned zone, respectively. Because the cell temperature $T_P(\vec{x}, \varphi)$ and most of the other variables at the central node (e.g. the density) are representative neither for the unburned zone nor for the burned zone, it is a remarkable cause of inaccuracy to use them in 3D-CFD-models which, e.g. expressively require information about the unburned zone (laminar flame speed, diesel-self-ignition, local burn rate, etc.) or about the burned zone (e.g. modeling of the burned gas properties). The example reported in Figure 9.11 helps understanding the influence of the cell dimension, the structure or even the orientation on the flame propagation.

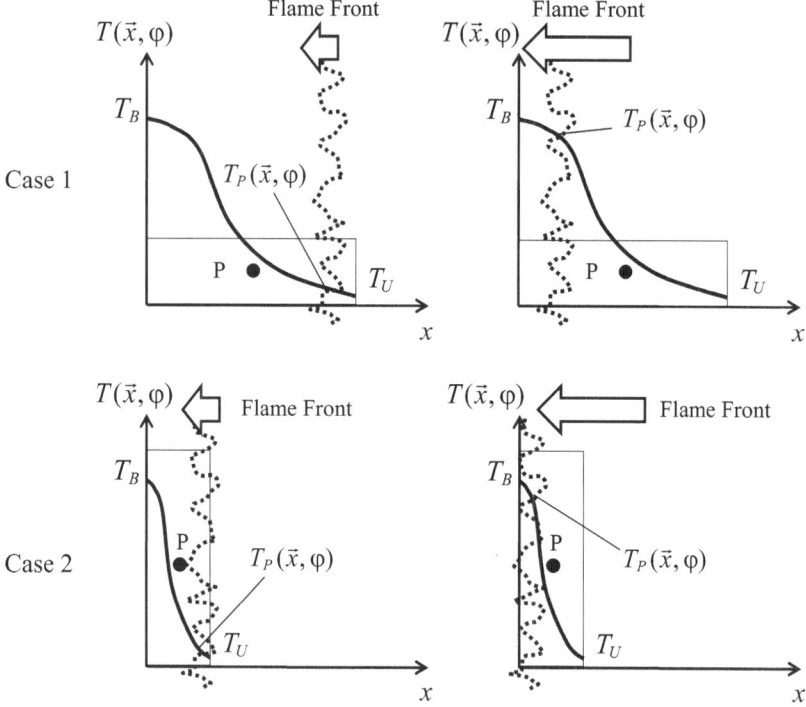

Figure 9.11: *Flame propagation through cells with different orientation.*

Case 1

In case 1 the heat release of the flame "heats" the temperature $T_P(\bar{x}, \varphi)$ of the cell very slowly from the start value $(T_P(\bar{x}, \varphi) = T_U(\bar{x}, \varphi))$. As soon as the temperature $T_P(\bar{x}, \varphi)$ rises, the laminar flame speed S_L "irrationally" increases and the flame propagation velocity due to this numerical problem accelerates up to its inconsistent maximum when the flame leaves the cell ($T_P(\bar{x}, \varphi) = T_B(\bar{x}, \varphi)$).

Case 2

Here the cell has the same dimension as in case 1 but a different orientation. In this case the flame heat release increases the cell temperature $T_P(\bar{x}, \varphi)$ with higher gradients. The resulting flame propagation speed S_L is "irrationally" many times higher than in case 1. Within a short time the flame leaves the cell while in case 1, during the same period, it has propagated through a smaller distance at the beginning.

Conclusion

Concluding, this means that the calculation of the flame speed using the only available temperature in the cell $T_P(\bar{x}, \varphi)$ is formally wrong and cannot be adjusted with a corrector factor, because numerically all depends on the "heating speed" of each cell, which is a result of complex non linear equations influenced by the position of the cell vertices, the flow field, the flame propagation direction, the gradients of the progress variables, etc. In particular considering the combustion chamber mesh of an internal combustion engine this has, in order to reproduce the complex geometry and the piston motion, moving cells with a great spectrum of dimensions, internal angles, etc., which actually can be taken into account only within a statistical approach.

9.3.2.1   Expedients for the Numerical Inconsistencies at the Flame Front

Since the local thermodynamic variables within a cell of the mesh are not reliable for a correct implementation of the flame propagation model an expedient that allows to handle the problem has to be found, e.g. referring to the temperature in the unburned zone at the flame front the following approaches may be an easy and reasonable solution of the problem:

- the average temperature of the whole unburned zone is taken as reference temperature (this approach misses local details)

- the average temperature of the neighbor unburned cells is taken as reference temperature (This approach is geometrically very complex and can be inconsistent in many cases.)

During the years and several investigations a more comprehensive and numerical convenient approach has been chosen. This approach is presented here below.

9.3.3 Local Two-Zones Model

In order to avoid the inaccuracy due to a wrong detection of temperature gradients described in Chapter 9.3.2, several years ago [66] a so-called "Local CFD-Two-Zones-Model" has been developed and implemented into *QuickSim*. This model permits a continuously thermodynamic splitting of each cell j involved in a combustion process into a burned and unburned zone with a similar procedure like in the real working-process analysis for the whole combustion chamber (see Figure 9.12). Following this approach also the local volumes of the two zones can be detected so that a more reliable evaluation of the flame front can be performed. Using the conservation equations for the mass and the energy within the cell j and a simple assumption for the relation between the unburned $\rho_{U,j}$ and burned density $\rho_{B,j}$, the equation system can be closed and a reliable thermodynamic splitting of all the cells involved in the oxidation process can be achieved.

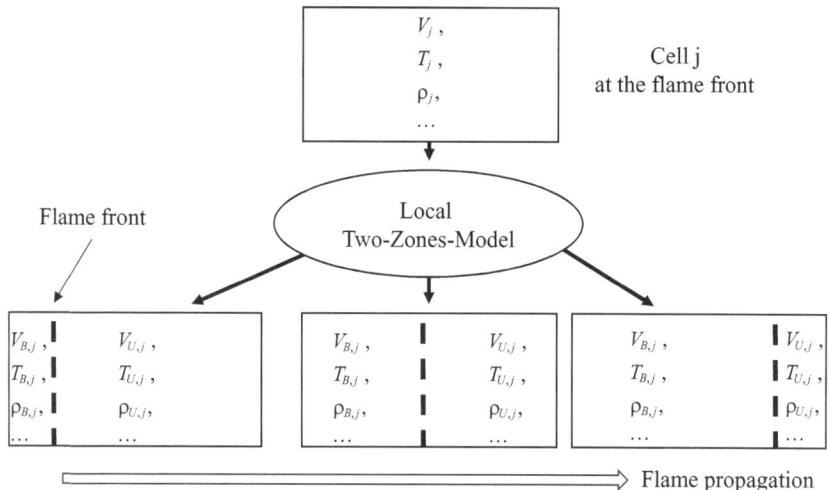

Figure 9.12: *Scheme of the "thermal splitting" of a flame front cell using a local two-zones model.*

The procedure starts with the following relations:

$$V_j = V_{U,j} + V_{B,j}$$ (9.17)

$$m_j = m_{U,j} + m_{B,j}$$ (9.18)

$$\rho_j \cdot V_j = \rho_{U,j} \cdot V_{U,j} + \rho_{B,j} \cdot V_{B,j}$$ (9.19)

$$m_j \cdot h_j = m_{U,j} \cdot h_{U,j} + m_{B,j} \cdot h_{B,j}$$ (9.20)

$$m_j \cdot \bar{c}_{p,j} \cdot T_j = m_{U,j} \cdot \bar{c}_{p,U,j} \cdot T_{U,j} + m_{B,j} \cdot \bar{c}_{p,B,j} \cdot T_{B,j}$$ (9.21)

where the mean specific heat $\bar{c}_p$, defined as follows, can be directly calculated from the thermal enthalpy available in *QuickSim* (see Chapter 8.4.2):

$$\bar{c}_p(T,p,\lambda) = \frac{\int_{T_{ref}}^{T} c_p(T,p,\lambda) \cdot dT}{T - T_{ref}} = \frac{h(T,p,\lambda)}{T - T_{ref}}$$ (9.22)

In a 3D-CFD-simulation it is more convenient to calculate with specific variables, i.e. the progress variable $w_{B,j}$ can be used in these equations:

$$w_{B,j} = \frac{m_{B,j}}{m_j}$$ (9.23)

so that the equations for the local unburned $T_{U,j}$ and burned temperature $T_{B,j}$ are given by:

$$T_{U,j} = \frac{\bar{c}_{p,j} \cdot T_j - w_{B,j} \cdot \bar{c}_{p,B,j} \cdot T_{B,j}}{(1 - w_{B,j}) \cdot \bar{c}_{p,U,j}}$$ (9.24)

$$T_{B,j} = \frac{\bar{c}_{p,j} \cdot T_j - (1 - w_{B,j}) \cdot \bar{c}_{p,U,j} \cdot T_{U,j}}{w_{B,j} \cdot \bar{c}_{p,B,j}}.$$ (9.25)

Considering the average densities over the entire combustion chamber and the global two-zones (e.g. in Figure 9.13 and 9.14 from *QuickSim*'s evaluation tool) the density profiles show a relatively constant ratio during the whole combustion duration over widespread applications. From this analysis and extending these considerations from the "global" zones to the local cell j it is possible to introduce an assumption that links the local unburned and burned density, respectively:

$$\rho_{U,j} = f(\rho_{B,j}) \cong 3 \cdot \rho_{B,j}$$ (9.26)

so that this relation in combination with the equation of ideal gas (assuming constant pressure within the cell) allows closing the equation system.

$$\rho_{U,j} \cdot R_{U,j} \cdot T_{U,j} = \rho_{B,j} \cdot R_{B,j} \cdot T_{B,j} . \tag{9.27}$$

It is important to underline the iterative aspect of this numerical procedure. At the first iteration in which a cell j is reached by the flame front, $T_{U,j}$ and $T_{B,j}$ are either assumed to be equal to the temperature of the central node T_j or they can be estimated; thus, also the specific heats $\bar{c}_p$ and the real gas constants R are evaluated at these initial temperatures. From these starting values (iteration 0), applying Eqs. 9.24, 9.25 and 9.27, the new temperatures of the two zones are evaluated and from these the values $\bar{c}_p$ and R are calculated (iteration 1). After few iterations the equation system converges to a plausible solution (see Figure 9.15).

Moreover, also the adiabatic flame temperature $T_{B\_ad,j}$ in cell j is determined using the local fluid properties in the unburned zone of the cell. During the calculation a comparison between the effective burned gas temperature $T_{B,j}$ and the adiabatic flame temperature $T_{B\_ad,j}$ (as a maximum value) is performed, so that, if necessary the variable $T_{B,j}$ can be numerically limited to $T_{B\_ad,j}$. Nevertheless, during many years of investigation it has been seen that $T_{B,j}$ remains always at least 10 K under the value of the adiabatic flame temperature.

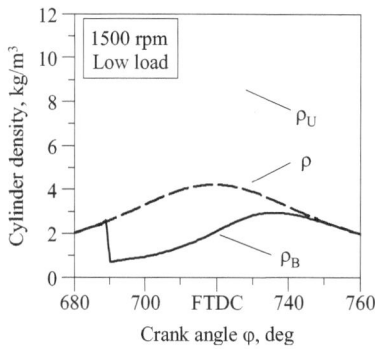
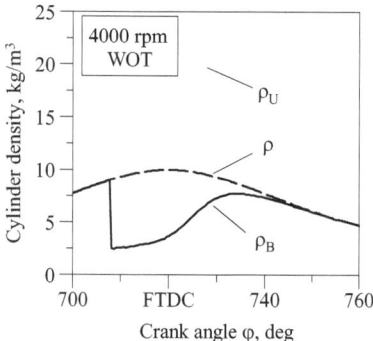

Figure 9.13: Unburned and burned density – (MB M102-E23)- 1500 rpm - low load – imep = 3.5 bar.

Figure 9.14: Unburned and burned density – (MB M102-E23) - 4000 rpm - WOT – imep = 10.2 bar.

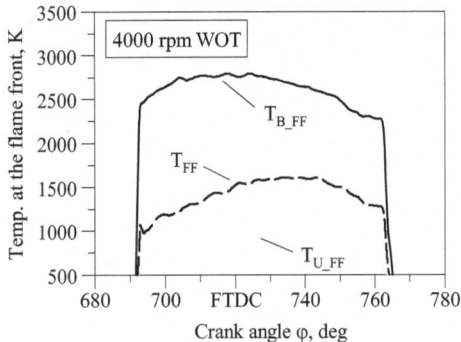

Figure 9.15: *Average temperature of the cells at the flame front (Rotax engine) – 4000 rpm – WOT – imep 10.3 bar.*

In Figure 9.15 the profiles of the temperatures as an average over all the cells at the flame front are reported. As it can be seen, the results of the presented local two-zones-model are plausible.

Comparison: Standard Approach vs. Local Two-Zones Approach

The following comparisons, using the same combustion model, show the differences in the profiles of the heat-release corrector factor $c_{HR}(\varphi)$ (see Chapter 9.3.1) with input variables either from the standard implementation or from the local two-zones-model. In the standard implementation the required values are provided by the central node of each cell involved in the process while in case of the local two-zones-model the required values are conveniently taken from either the unburned or the burned zones.

The results of two engine simulations at two completely different operating conditions (see Figures 9.16 and 9.17) show that in case of the two-zones-model the factor $c_{HR}(\varphi)$ remains roughly constant during the combustion process. Also the profile of $c_{HR}(\varphi)$ is quite similar and approximately $c_{HR}(\varphi) \cong 1$ in comparison between the presented simulations.

In case of the standard implementation in contrast, $c_{HR}(\varphi)$ shows much wider variations during the combustion process and also from simulation to simulation, i.e. in this case an overall correction factor, to be implemented in the combustion model, would not have a general validity.

Using flame propagation models, the calibration of the heat-release rate, independent from the implementation procedure, is usually performed by the help of a corrector factor A_{comb} applied to the turbulence velocity to be introduced in Eq. 9.7:

$$\overline{v}_T = \left(\left| \overline{v} \right|^2 + A_{comb} \cdot u'^{\,2} \right)^{1/2} \tag{9.28}$$

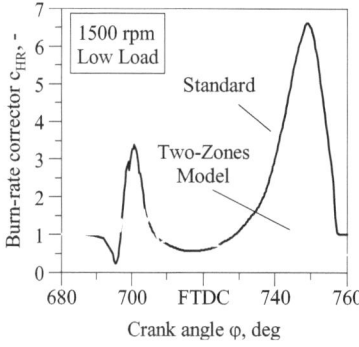

Figure 9.16: *Heat-release corrector factor $c_{HR}(\varphi)$. (MB M102-E23) 1500 rpm - low load – imep = 3.5 bar.*

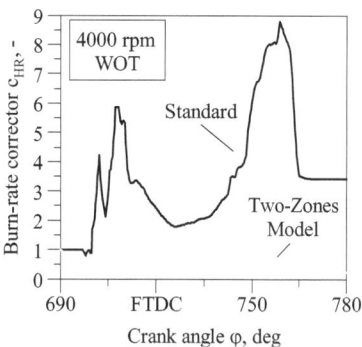

Figure 9.17: *Heat-release corrector factor $c_{HR}(\varphi)$. (Rotax engine) 4000 rpm - WOT – imep = 10.3 bar.*

Of course, also using the standard implementation procedure, it is practically always possible to fix a reasonable calibration constant A_{comb}. This constant allows a matching, for a given mesh and a more or less wide range of operating conditions, at least for the combustion duration and the position of 50% burned mass. The determination of A_{comb} is actually a compromising solution for all local inaccuracies at the flame front, but often it does not permit a sufficient predictability of the combustion model out of the calibration range. In contrast the local two-zones-approach allows the following enhancements:

- minor effort in the calibration of the combustion model

- a more reliable global and local flame propagation (less mesh dependence)

- a better predictability also for complex combustion strategies (e.g. stratified mixture)

- a more reliable description of the flame front is available also for other 3D-CFD-models (e.g. local self ignition models).

Since very often a quite irregular profile of the corrector factor $c_{HR}(\varphi)$ remains in the first stage of the flame development after the ignition (see Figures 9.16 and 9.17), during the years, an improvement towards a more reliable prediction during this combustion phase has been achieved. Below the concept idea of this improvement is reported.

Also at the end of the combustion the corrector factor $c_{HR}(\varphi)$ shows some irregularities. Here the remaining fresh charge mass is very small and the burn rate is low (see Figure 9.9) so that the effects on the combustion profile are much less sensitive then at the beginning of the flame propagation. Further models for a better calculation of the flame propagation within the cells in the wall-near region (mainly responsible for the profile of $c_{HR}(\varphi)$ at the end of the combustion) are already in the development phase.

9.3.4   Ignition Model

After the spark ignition the flame starts propagating from an initial flame kernel generated by the energy release of the electric arc (plasma region). At the beginning the flame propagates with a relative laminar flame speed S_L through the unburned zone and when the flame front has reached a dimension comparable to the turbulent eddies it accelerates to the turbulent flame speed S_T. As introduced in Chapter 9.3.3 (see Figures 9.16 and 9.17), independently of the local discretization degree of the mesh, the flame is always numerically very sensitive. The flame starts propagating within few cells and for a while (ca. 10 deg) it has a dimension not comparable to the average discretization length of the cells in the spark plug region. Therefore the prerequisites for a reliable calculation of the local flow field and consequently the flame propagation are not given.

Up to a certain radius r_K (usually $r_K \cong 4\text{-}5$ mm) the flame front can be assumed spherical with good accuracy, thus it is convenient to introduce a phenomenological approach based on a quasi-dimensional model for the real working-process analysis in this early flame development. As mentioned in Chapter 4.4.3.2 the main limitations of these quasi-dimensional models are in the determination of the flame front shape when, reaching a considerable dimension (usually $r_K \geq 8\text{-}10$ mm), the flame interacts with the walls of the combustion chamber. That means that as long as the flame propagates either spherically or half spherically the predictability of the quasi-dimensional model, in particular in this case where relevant input variables can be detected from the 3D-CFD-simulation, is very good.

Similarly to the wall heat-transfer calculation (see Chapter 10.2.5) an "internal coupling" between the 3D-CFD-simulation and a phenomenological quasi-dimensional model is established (see Figure 9.18). This procedure, as a "closed loop", allows to control the flame propagation in the 3D-CFD-simulation using a phenomenological spherical kernel growth with a given flame speed (laminar and turbulent) as a target value.

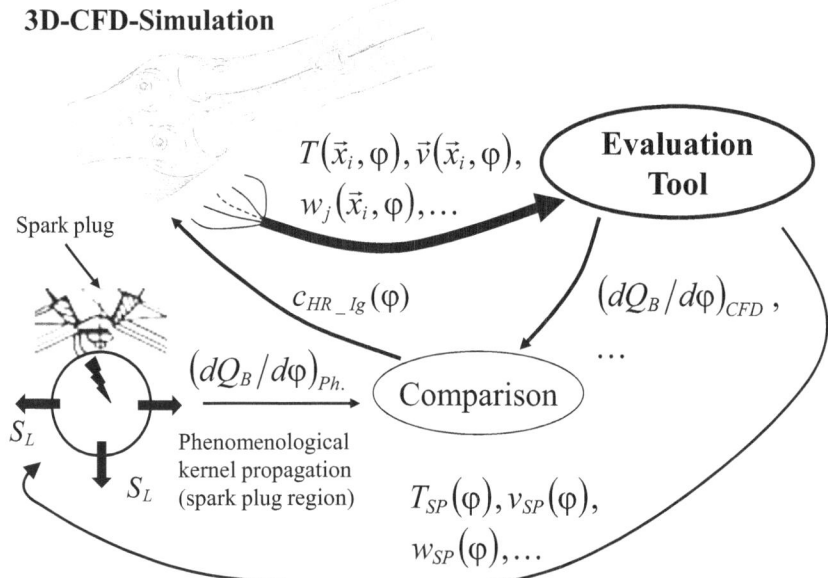

3D-CFD-Simulation

Figure 9.18: *The "internal coupling" between a 3D-CFD-code and a phenomenological quasi-dimensional model for early flame propagation.*

The control is performed by a correction factor $c_{HR\_Ig}(\varphi)$ (see Figure 9.18) that adjusts the flame propagation at each time-step up to $r_K \cong 4\text{-}5$ mm (after reaching this radius the combustion model switches to the common procedure). The controlling relation (in this case using as output variable the related heat-release) becomes:

$$c_{HR\_Ig}(\varphi) = \left(\frac{dQ_B}{d\varphi}\right)_{Ph.}\bigg/\left(\frac{dQ_B}{d\varphi}\right)_{CFD}.$$

(9.29)

Starting with the definition of the flame kernel volume $V_{B\_Ig}(\varphi)$ as a sum of the local burned zone of the involved cells ($j = 1\ldots N_{B\_Ig.}$)

$$V_{B\_Ig} = \frac{4\cdot\pi\cdot r_K^3}{3} = \sum_{j=1}^{N_{B\_Ig.}} \frac{w_{B,j}\cdot\left(\rho_{U,j}/\rho_{B,j}\right)\cdot V_j}{1 + w_{B,j}\cdot\left(\rho_{U,j}/\rho_{B,j}\right) - w_{B,j}}$$

(9.30)

and applying the two-zones-model presented in Chapter 9.3.3, it is then possible to calculate the target heat release for a flame propagation with a relative flame speed S_L or S_T. The specific heat release term Δh_{HR} is calculated according Chapter 8.4.2.1:

$$\left(\frac{dQ_B}{d\varphi}\right)_{Ph} = \rho_U \cdot A_f \cdot S_{L/T} \cdot \Delta h_{HR} = 4 \cdot \pi \cdot r_K{}^2 \cdot \rho_U \cdot S_{L/T} \cdot \Delta h_{HR} \,. \tag{9.31}$$

Finally it is possible also to calculate the absolute flame speed u_f

$$u_f = \frac{\partial V_{B\_lg}/\partial t}{A_f} = \frac{\partial V_{B\_lg}/\partial t}{4 \cdot \pi \cdot r_K{}^2} \tag{9.32}$$

so that a comprehensive evaluation and control of the flame propagation during the early flame development can be performed.

9.3.5 Final Implementation Procedure

The combustion model presented in this chapter is based on the Weller formulation. The implementation within the 3D-CFD-code *QuickSim* has been performed choosing input variables able to give reliability to the fuel heat release over a wide range of engine operating conditions and trying to reduce as much as possible the dependence of the mesh on the results.

In this section the numerical implementation procedure of each variable is briefly described..

Laminar Flame Speed

The laminar flame speed S_L according to the proposed combustion model is the basic variable for the calculation of the final flame propagation. Starting with the formulation introduced with Eqs. 4.17 and 4.18, the local calculation of S_L in a cell j within the flame front takes the following input variables into account:

$$S_L = f(\lambda_{Fresh,j}, p, T_{U,j}, w_{EGR\_U,j}) \tag{9.33}$$

where $\lambda_{Fresh,j}$ and $\lambda_{EGR\_U,j}$ are easily obtained using the local scalar values of the species (see Chapter 8.4). The local temperature of the unburned zone $T_{U,j}$ is provided by the two-zones-model introduced in this chapter. Since the pressure p in the cylinder is almost constant during the working period, there is no need to use the local value p_j, i.e. the spatial-average pressure p is directly provided to the calculation model of S_L by the evaluation tool of *QuickSim* (see Chapters 7.4 and 10.2.4).

Turbulent Term

In the formulation of the turbulent term (see Eq. 9.28) the local fluid velocity in each cell at the flame front is taken as a reference for the variable v.

Regarding the description of the turbulent velocity u', the procedure is more complex. As already mentioned in this work the turbulence is a very sensitive variable with the mesh quality (more details in Chapter 10.2.4.1), i.e. the local value of the turbulence can often be an "unreliable" input variable.

The problem can be pragmatically solved using a filtered result of the turbulence. Therefore in the cells at the flame front the relevant turbulence value u' for the combustion model is calculated using a weighting factor C_{turb} that mixes the local turbulent velocity u'_{local} with the average value over the whole unburned zone u'_U (see Figure 9.19).

$$u' = C_{turb} \cdot u'_{local} + (1 - C_{turb}) \cdot u'_U \qquad (9.34)$$

Depending on an estimation of the mesh quality during the whole simulation (the evaluation tool of *QuickSim* automatically provides mesh quality indexes, like average aspect ratio, warpage angle, etc.) the factor C_{turb} usually varies between 0.5 and 0.7. This factor can also be set selectively in the different regions of the combustion chamber. Exemplarily in the cells in the near-wall region, where the turbulence is extremely sensible with the mesh quality, it is often convenient to use locally very low values of C_{turb} (~0.1-0.3).

Since a too high value of the turbulence may cause a local flame extinction (see Chapter 9.1), if the ratio u'/S_L exceeds critical values provided by the internal flame quenching models, the local burn rate can be either reduced or completely deactivated.

Turbulent Length Scale

A similar procedure adopted for the turbulence is used here also for the description of the relevant turbulent length scale l_l for the combustion model.

$$l_l = C_{tls} \cdot l_{l,local} + (1 - C_{tls}) \cdot l_{l,U} \qquad (9.35)$$

Therefore in the cells at the flame front l_l is calculated using a weighting factor C_{tls} that mixes the local turbulence length scale $l_{l,local}$ with the average value over the whole unburned zone $l_{l,U}$ (see Figure 9.20). The influence of the turbulent length scale on the local burn rate is less relevant then the turbulence, what allows the usage of relatively high values of C_{tls} (~0.7-0.9).

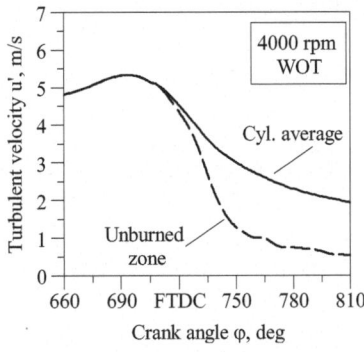

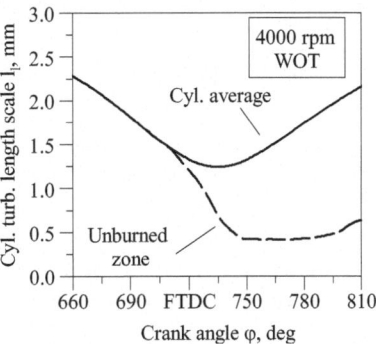

Figure 9.19: *Turbulent velocity – (MB M102-E23) - 4000 rpm - WOT – imep = 10.2 bar.*

Figure 9.20: *Turbulent length scale – (MB M102-E23) - 4000 rpm - WOT – imep = 10.2 bar.*

Flame Radius

The required flame radius r_K is at disposal from the evaluation tool of *QuickSim* and is directly implemented in the formulation of Weller's combustion model within a "closed loop".

9.4 Results

Figures 9.21 and 9.22 show the result of the cylinder pressure profile in the Rotax engine using the proposed implementation of the combustion model without additional adjustments for the different operating conditions (i.e. the same A_{comb} has been used).

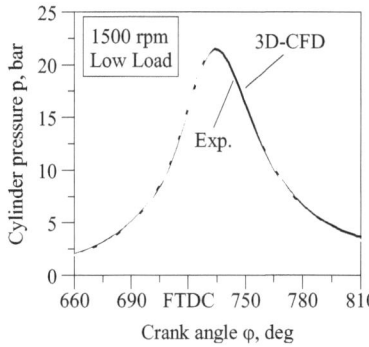

Figure 9.21: *Cylinder pressure.*
Comparison 3D-Ph. vs. Exp. - (Rotax)
1500 rpm - low load – imep = 4.0 bar.

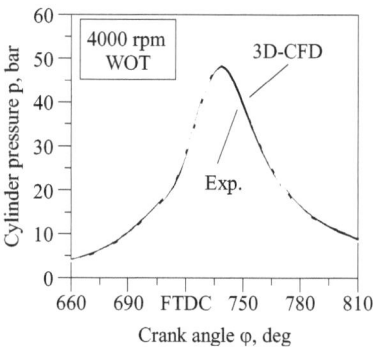

Figure 9.22: *Cylinder pressure.*
Comparison 3D-Ph. vs. Exp. - (Rotax)
4000 rpm - WOT – imep = 10.3 bar.

The comparisons with experimental data show a very good accuracy at both operating conditions, which remarkably differ in terms of engine speed and load.

10

3D-CFD-Modeling of the Wall Heat-Transfer

The wall heat-transfer represents the thermodynamic boundary condition of the combustion chamber and, in particular during the working period, is the main source of inaccuracy in the 3D-CFD-analysis of the engine operating cycle.

Very often in the 3D-CFD-simulation attention is mainly paid to the modeling of phenomena like combustion (chemical kinetic reactions) and fuel spray formation, instead of ensuring an accurate balance in the solution of all relevant engine processes. This is a fatal error that easily compromises the quality of the simulation results because a simulation running with wrong thermodynamic boundary conditions will never deliver reliable results.

10.1 Introduction

During the last 50 years the heat-transfer process between the working fluid and the combustion chamber walls and its noticeable effects on engine performance, efficiency and emissions have been intensively experimentally investigated and reported. The heat transfer is one of the major factors affecting the engine energy balance at any operating condition. Considering exemplary a common natural aspirated SI engine [5,7,26,40,71-86] (see Figure 3.2), the ratio of heat transfer to the fuel heating energy is approximately 20 – 25 % for full load and even more than 30 % for low load, respectively. Therefore, a high accuracy in the determination of the engine heat transfer is a prerequisite for a realistic computational analysis according, at least, to the engine energy balance.

In comparison to other engine processes, whose evaluation can be easily proved with reference values (e.g. during the combustion the heat release will never exceed 100% of the fuel heat-value) the "plausibility check" of the amount of heat-transfer losses is in fact limited to a comparison of the before introduced heat-transfer to heat-release ratios with range values according to experience.

10.1.1 Phenomena Understanding, Calculation Approach and Considerations

The convective heat-transfer process is governed by the temperature and velocity profile within the near-wall region (boundary layer), which is, in internal combustion engines, locally and temporally drastically under the influence of the mean flow motion, the turbulence, the flame propagation, etc. – *limited physical understanding* -.

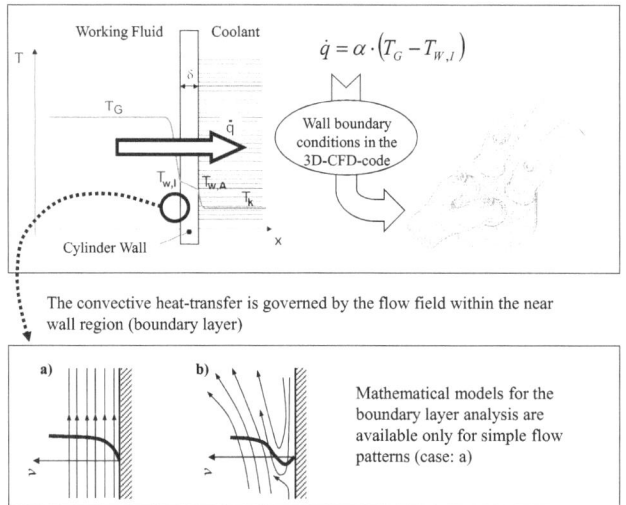

Figure 10.1: The transient boundary layer.

In order to solve the Navier-Stokes conservation equations within the boundary layer (laminar sub-layer and turbulent boundary layer) complex mathematical models for anisotropic turbulence, turbulence separation, impinging jet, flame quenching phenomena, etc., are required – *limited mathematical formulation* -. Nowadays such mathematical models have been proposed only for simple wall geometries and flow patterns, respectively [37,54,56]. Therefore, not a single validated model, which permits an accurate investigation of the boundary layer of a combustion chamber, has been successfully implemented in simulation programs.

The above mentioned complexity in analyzing the near-wall region behavior explains the reason why the state-of-the-art calculation of the wall heat flux in the three-dimensional simulation represents the major source of inaccuracy. In addition, the local mesh structure and its dimensions, especially in the wall-cells layer, drastically influence the results of local as well as overall heat transfer – *limited numerical implementation* -.

Few years ago the opinion was not rare that by means of the detailed approach of the 3D-CFD-simulation semi-empirical heat-transfer relations developed for the thermodynamic real working-process analysis could be verified and improved significantly. In the meantime it has been found that the wall functions implemented as a standard in 3D-CFD codes underestimate the wall heat losses in the combustion chamber by a factor of 5 to 10 [10,11,21,22], what was proven by measurements. Why an explicit heat-transfer calculation is largely impossible within the scope of a 3D-CFD-simulation is shown in Figure 10.2.

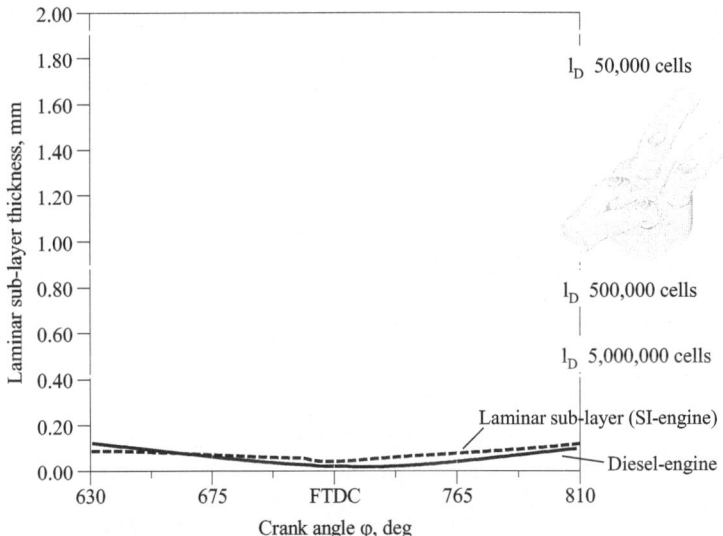

Figure 10.2: Cell discretization l_D and the laminar boundary sub-layer.

Using a simple relationship, the thickness of the laminar sub-layer of the thermal boundary layer can be calculated from the heat-transfer coefficients [10,37], where heat flux to the combustion chamber wall takes place by heat conduction only. If the profile of the laminar sub-layer is compared to the averaged cell sizes with varying refinement degrees of the combustion chamber, it becomes obvious that even with 5,000,000 cells (only for the combustion chamber) there is no possibility of discretizing the thermal boundary layer. The required grid refinement - even if only in the immediate proximity to the wall - would lead to the problem that the greatest part of the entire computing time must be used for calculating the wall heat-transfer.

In addition to limitations in the physical understanding, the mathematical formulation and the numerical implementation described before, there is another variable that remarkably affects the heat-transfer process: the wall temperature. The wall temperature depends not only on the processes at the cylinder gas-side but also on the cooling effects at the cylinder water-side. As well known its experimental determination is very time and cost expensive and possible, if at all, only in some parts of the combustion chamber surface (cylinder liner and cylinder head). These experimental investigations are usually dedicated to research engines so that practically the local wall temperatures for simulation purposes can be only estimated.

Simple "wall temperature models" [71-74] based on heat-capacities of all engine components can help estimating the average wall temperature of each component and in few cases also the distribution (e.g. cylinder liner and piston crown). The future improvement of simulation programs based on the coupling of 3D-CFD and FEM approaches, represent a promising way towards a satisfactory reliability and acceptable time-expenses for a more accurate determination of the wall temperature distribution.

10.2   State-of the-Art of Engine Heat-Transfer Calculation in the 3D-CFD-Simualtion

The state-of-the-art calculation of the heat-transfer coefficient is commonly based on the standard "Wall Function Approach". Furthermore several LRN-Models (Low Reynolds Number) are rarely implemented. As mentioned before the state-of-the-art calculation is still not satisfactory and future developments of 3D-CFD-heat-transfer-models based on the local analysis of the boundary layer seem to be less promising.

For this reason a new approach based on the formulation for the working-process analysis (phenomenological approach) is introduced and analyzed.

10.2.1 The Wall Function Approach

In the wall function approach, the flow and temperature fields within the boundary layer as well as the local heat-transfer coefficient $\alpha(\bar{x}, \varphi)$ are obtained by algebraic relations based on a "particular predicted" distribution of velocity, temperature and turbulence parameters. Practically, the aim of the wall function approach is to bridge the solution between the turbulence transport equations in the interior flow where the turbulence is fully developed (high Reynolds number domain) and the no-slip condition imposed at the wall. The prediction of the boundary layer field is usually obtained by the following main assumptions (steady-flow over a flat plate) [54]:

- The velocity, temperature, etc. vary predominantly normal to the wall (one-dimensional behavior).

- Pressure gradients and body forces are neglected (uniform shear stress in the boundary layer).

- Shear stress and velocity are aligned and unidirectional throughout the layer.

- The production of turbulence energy and its dissipation are balanced.

- There is a linear variation of the turbulence length scale.

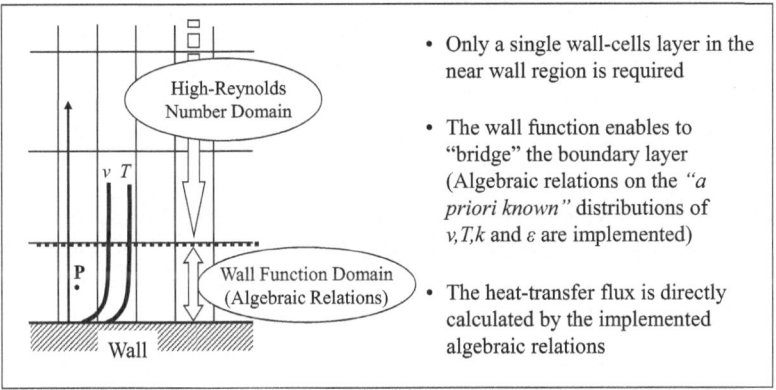

Figure 10.3: *The wall function approach in the near-wall region (boundary layer).*

The numerical implementation of the wall function (see Figure 10.3) requires a single wall-cells layer, whose central node P is assumed to be placed outside the boundary layer. Here the introduction of a dimensionless wall distance y^+ helps investigating the distance to the wall y where the turbulence becomes isotropic. This variable, similarly to the definition of the Reynolds number Re, describes the competition between the shear velocity u_τ (component of the vector $\bar{v}$ parallel to the wall) and dumping processes caused by viscous dissipation ($\sim$ to dynamic viscosity μ).

$$y^+ = \frac{\rho \cdot u_\tau \cdot y}{\mu} \tag{10.1}$$

The optimal value of the dimensionless normal distance y^+ from the wall should be approximately 30-100 units [56]. If the central node P is positioned too close to the wall the utility of the wall function is practically invalidated. This approach is also an expedient because it avoids the need to set a fine computational mesh within the boundary layer with associated computing overheads.

The accuracy depends on both the fulfilling of the y^+-condition and on the degree to which the assumptions and approximations embodied in this approach correspond with the reality of the application. The flow in an internal combustion engine completely differs from the simple assumptions listed above. In addition, the complexity in optimizing the mesh motion, especially at TDC, due to the high compression degree of the cell layers, does not permit to satisfy the y^+-condition at many locations.

10.2.2 Low Reynolds Number Models

In this approach, the conservation equations within the boundary layer are solved by using complex models for anisotropic turbulence (LRN-Models), with a no-slip condition imposed at the boundary cell faces. At some distance from the wall, a switch is made from the high-Reynolds-Number model to the chosen LRN-Model.

In contrast to the wall function, the numerical implementation requires a fine mesh of at least 15 cell-layers in the near-wall region, up to a thickness sufficient to encompass the boundary layer (see Figure 10.4) [56]. However, since its thickness is not known a priori, several trial-and-error adjustments concerning the mesh refinement in the near-wall region are necessary.

As introduced before the required discretization for the implementation of LRN models is first of all still extremely coarse in comparison to the characteristic scale of the boundary layer. This leads a priori to an inaccurate analysis of the fluid motion in the near-wall region. In addition detailed inputs like wall temperature and surface properties (oil, soot formation, etc.) are missing and the difficulty to ensure an adequate mesh motion and structural mesh quality with so many additional layers near the wall at any piston position is extremely high.

At the end the implementation of LRN-Models in the simulation of internal combustion engines causes remarkable computing overheads (3-4 times the CPU-time of an equivalent mesh using the wall function approach) and despite this effort the results of some simulations using LRN-Models [75] show no considerable improvements in calculating the engine heat transfer.

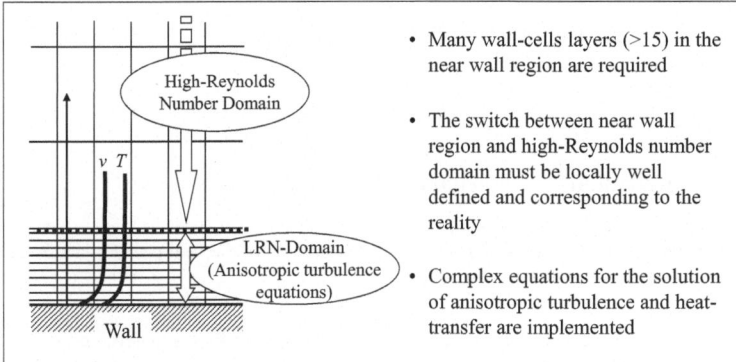

Figure 10.4: *The LRN-approach in the near-wall region (boundary layer).*

10.2.3 Phenomenological Heat-Transfer Models in the Real Working-Process Analysis (WP)

Since the end of the sixties, phenomenological heat-transfer models have been developed and implemented in the real working-process analysis (WP). For that application the instantaneous spatially averaged convective heat-transfer coefficient $\alpha(\varphi)$ is required [25,36,39]. Therefore, the heat transfer to the combustion chamber surfaces in contact with the unburned and burned gas zones is given by:

$$\frac{dQ_W}{d\varphi} = \alpha(\varphi) A_W(\varphi) \cdot [T_G(\varphi) - T_W] \frac{dt}{d\varphi} \tag{10.2}$$

The phenomenological approach is based on the Re-Nu-correlation (dimensional analysis), in which the assumption of the Nusselt, Reynolds and Prandtl numbers relationships follow that found for turbulent flow in pipes:

$$Nu = \frac{\alpha(\varphi) \cdot l}{\lambda} = a \cdot \mathrm{Re}^m \cdot \mathrm{Pr}^n \tag{10.3}$$

The following exponent values, $m = 0.78$ and $n = 0.33$, have been found out as representative for the engine heat-transfer process [26,39,40]. In the usual temperature and pressure range of internal combustion engines the Prandtl number Pr can be assumed as a constant $(\mathrm{Pr} = 0.7)$. The value Pr^n is then contained in the constant a. The coefficient $\alpha(\varphi)$ becomes (from Eq. 10.3):

$$\alpha(\varphi) = a \cdot l^{-0.22} \cdot \lambda(\overline{T}) \cdot \left(\frac{w(\varphi) \cdot \rho(\overline{T}, p)}{\mu(\overline{T})} \right)^{0.78} \tag{10.4}$$

This equation represents the basic formulation of the phenomenological heat transfer model.

10.2.3.1 Motivation for a Phenomenological Approach

The motivation for a phenomenological approach in modeling the heat transfer can be resumed as follows:

> *The greatest advantage using the dimensional analysis is to develop*
> *a functional form of relationships which govern a special process mechanism.*
> *Therefore, even in an extremely complex phenomenon, like the engine heat-transfer,*
> *it is possible to identify a limited number of critical variables,*
> *which are predominately responsible for the process.*

Using these phenomenological heat transfer models, the critical choices to be made are:

- Characteristic length l of the combustion chamber geometry.

- Characteristic velocity w of the fluid motion.

- Relevant gas temperature $\overline{T}$ at which the gas properties are evaluated.

Different procedures and the basis of the derivation of the characteristic variables have been proposed and experimentally validated by several authors. The most widely used correlations are, chronologically reported, from Woschni [39,76], Hohenberg [40] and Bargende [26].

10.2.3.2 Woschni's Correlation

Woschni's correlation was first formulated for diesel engines only, but it has been proved to calculate the heat-transfer in spark ignition engines with a comparable accuracy. Briefly described, this correlation takes the cylinder bore B as the characteristic length and the cylinder-average charge temperature T_G as the relevant gas temperature [7,26,39]:

$$\alpha(\varphi) = 130 \cdot B^{-0.2} \cdot p^{0.8} \cdot T_G^{-0.53} \cdot \left(w \right)^{0.8} \tag{10.5}$$

where the velocity term w, which is interpreted as the averaged cylinder gas velocity, is expressed as follows:

$$w = \left[C_1 \bar{S}_P + C_2 \frac{V_D \cdot T_{G,IVC}}{p_{IVC} \cdot V_{IVC}} \cdot (p - p_m) \right] \qquad (10.6)$$

($p_{IVC}, V_{IVC}, T_{G,IVC}$ are the gas pressure, the volume and the temperature at intake valve closing: IVC). C_1 and C_2 are constants, which assume different values during each stroke. The second part of the velocity term $(\propto (p - p_m))$ embodies the effect of the combustion process on the engine heat transfer.

10.2.3.3 Hohenberg's Correlation

Hohenberg's correlation was developed especially for a better prediction of the heat-transfer at low-load engine operating conditions. In this formulation, instead of the bore, the diameter $\overline{D}$ of a sphere of the same volume like the instantaneous cylinder volume V is taken as the characteristic length [40]:

$$\overline{D}^{-0.2} = C \cdot \left(V^{0.33} \right)^{-0.2} \quad \Rightarrow \quad \alpha(\varphi) = 130 \cdot V^{-0.06} \cdot p^{0.8} \cdot T_G^{-0.53} \cdot \left(w \right)^{0.8} \qquad (10.7)$$

where the velocity term w includes the effect of the gas temperature T_G:

$$w = \left[T_G^{0.1625} \cdot \left(\bar{S}_P + 1.4 \right) \right] \qquad (10.8)$$

Hohenberg's correlation is then given by:

$$\alpha(\varphi) = 130 \cdot V^{-0.06} p^{0.8} T_G^{-0.4} \left(\bar{S}_P + 1.4 \right)^{0.8} \qquad (10.9)$$

The effect of the combustion process has not been included in the formulation. This correlation permits a good accuracy in predicting the heat-transfer during the gas exchange period, but it underestimates the heat losses during the working cycle [7,26].

10.2.3.4 Bargende's Correlation

At the beginning of the nineties, an improved correlation for the working cycle of spark ignition engines, that takes additional influencing factors into account, was published by Bargende [26]. Compared to the previous reported correlations, Bargende's basic formulation considers expressly the effect of the combustion process on the engine heat transfer introducing an additional term Δ:

$$\alpha(\varphi) = a \cdot \overline{D}^{-0.22} \cdot \lambda(\overline{T}) \cdot \left(\frac{w(\varphi) \cdot \rho(\overline{T}, p)}{\mu(\overline{T})} \right)^{0.78} \cdot \Delta$$ (10.10)

Here, like in Hohenberg's correlation, the diameter $\overline{D}$ of a sphere of the same volume like the instantaneous cylinder volume V is taken as the characteristic length. The relevant gas temperature $\overline{T}$ at which the gas properties are evaluated is, instead of T_G, the estimated average temperature within the boundary layer:

$$\overline{T} = \frac{T_G + T_W}{2}$$ (10.11)

The influence of turbulence on the engine heat transfer is included in the velocity term w. Instead of the mean piston velocity $\overline{S}_P$, the instantaneous piston velocity S_P is argued to be proportional to the average gas velocity in the cylinder:

$$w = \frac{1}{2} \cdot \sqrt{\frac{8}{3} \overline{k}_{0D} + S_P^2}$$ (10.12)

In this formulation the average turbulent kinetic energy $\overline{k}_{0D}$ is provided by a simple global zero- or quasi-dimensional $(\overline{k} - \overline{\varepsilon})_{0D}$ turbulence model (for details [7,12,26]).

Combustion Term

The main goal of the term Δ is to take the flame propagation process into account by dividing the combustion chamber into two zones: the unburned zone V_U and the burned zone V_B, respectively. For this purpose two sub-terms $[A] \propto V_B/V$ and $[B] \propto V_U/V$ have been introduced:

$$\Delta = f([A],[B])$$ (10.13)

with:

$$[A] = w_B \cdot \frac{T_B}{T_G} \cdot \frac{T_B - T_W}{T_G - T_W}$$ (10.14)

and:

$$[B] = (1 - w_B) \cdot \frac{T_U}{T_G} \cdot \frac{T_U - T_W}{T_G - T_W}$$ (10.15)

where w_B is the burned mass fraction in the cylinder. Bargende, at that time, found out empirically that the best agreement with numerous experimental investigations is achieved when:

$$\Delta = ([A] + [B])^2 \tag{10.16}$$

At the end, Bargende's correlation is then given by:

$$\alpha(\varphi) = 253.5 \cdot V^{-0.073} \cdot p^{0.78} \cdot \overline{T}^{-0.477} \cdot w^{0.78} \cdot \Delta \tag{10.17}$$

Many studies in the last decade show that this correlation obtains a high accuracy in predicting the engine heat-transfer, especially during the combustion process [12,41,71]. For about 20 years, Bargende's correlation has been implemented in modern high and low pressure indicating systems combined with software for the real working-process analysis. This system represents the avant-garde in the SI-engine combustion optimization, in which high accuracy in predicting the heat transfer is required.

10.2.4 Comparison between the 3D-CFD-Heat-Transfer (Wall-Function Model) and the Real Working-Process Analysis

This chapter shows some results of few different 3D-CFD-simulations using a state-of-the-art wall function heat-transfer model (WF) provided by a commercial 3D-CFD-code (Star-CD, Version 3.15). The calculated overall heat-transfer rate $dQ_W/d\varphi$ (J/deg) as well as the cumulative heat transfer energy on a cycle Q_W (J/cycle) have been compared with the results of the real working-process analysis (WP).

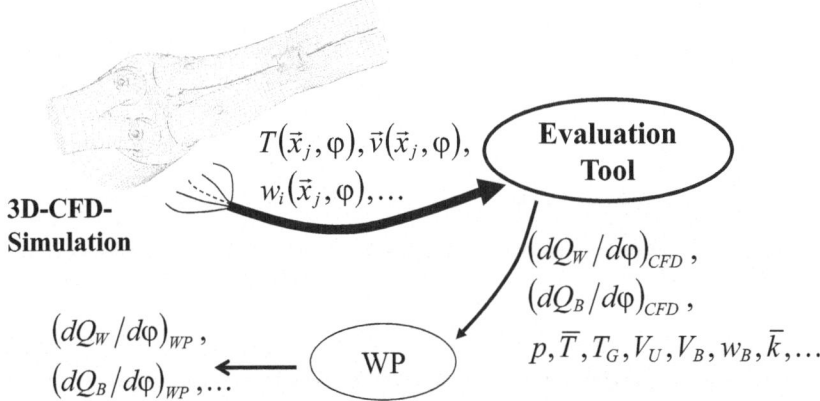

Figure 10.5: Heat transfer calculation: Two models at the same thermodynamic conditions.

In order to ensure the same "thermodynamic conditions" during the comparison (see Figure 10.5), the necessary average values of the variables in the real working-process analysis have been automatically provided by the evaluation tool implemented in *QuickSim* (see Chapter 7.4). In this work, the heat-transfer models chosen in the real working-process analysis are:

- Bargende's correlation during the compression and expansion stroke (working cycle).

- Hohenberg's correlation during the intake and exhaust stroke (gas exchange period).

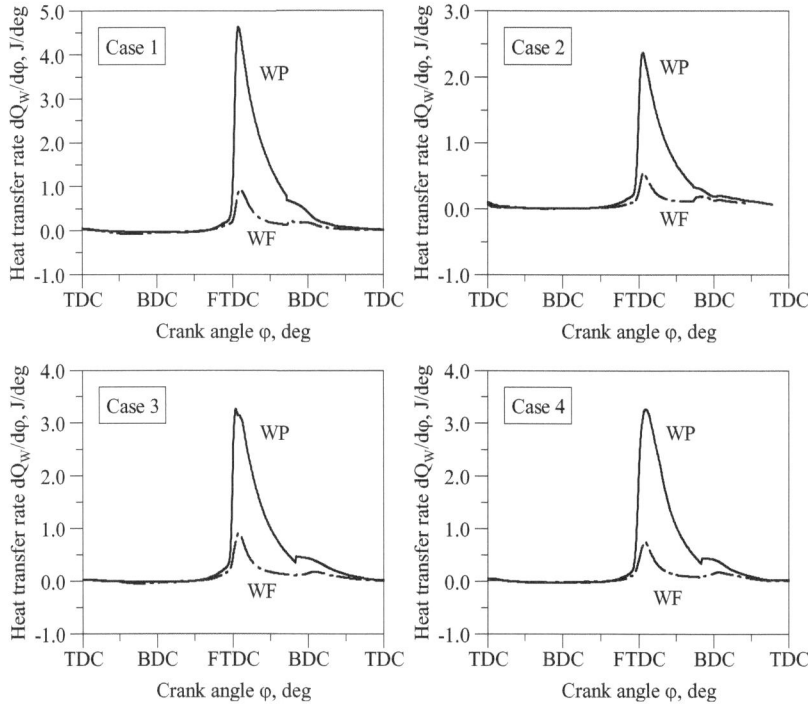

Figure 10.6: *Comparison between the 3D-CFD-heat-transfer (WF: wall-function model) and the real working-process analysis (WP) under the same "thermodynamic conditions".*

Here a representative selection of numerous comparisons is reported. In cases 1 und 2 (see Table 10.1) two operating conditions of an old mass production engine are reported. Case 3 and 4 refer to test engines with very simple combustion chamber geometries, so that it is possible to implement a 3D-CFD-mesh with a high structure quality (low warpage angle, optimal internal face angle, optimal aspect ratio and a better fulfilling of the y^+-condition [56]) at any piston position.

Table 10.1: Comparison: 3D-CFD wall heat-transfer calculation with wall function models (WF) vs. real working-process analysis (WP).

Specifications:	Case 1		Case 2		Case 3		Case 4	
Engine	MB M102-E23		MB M102-E23		Test-Engine		Test-Engine	
No. of valves/cyl.-	2		2		2		4	
Cyl. displacement, cm^3	575		575		480		480	
Combustion chamber shape	Bowl + squish area		Bowl + squish area		Pent-roof		Flat	
rpm,-	3,000		1,500		4,000		3,000	
Load, -	WOT		Low		WOT		WOT	
imep, bar	10.8		3.4		10.5		9.5	
3D-CFD-Mesh, N_{Cells}	15,000		15,000		45,000		50,000	
Cell type, -	Hexahedral		Hexahedral		Hexahedral		Hexahedral	
Simulated fuel, -	C_8H_{18}		C_8H_{18}		C_3H_8		C_8H_{18}	
	WF	WP	WF	WP	WF	WP	WF	WP
Heat release Q_B, J/cycle	1684	1727	621	619	1455	1455	1324	1311
Heat transfer Q_W, J/cycle	62.3	352.4	61.5	202.4	77.1	286.6	66.2	308.1
Ratio Q_W/Q_B , %	3.7	20.4	9.9	32.7	5.3	19.7	5.0	23.5

The four cases show that the heat-transfer prediction using the wall function is in any case not plausible. Due to the high mesh-structure quality, even in cases 3 and 4 no remarkable improvements have been achieved. These results confirm the analyses of other authors [11]. The reason of this inaccuracy, as explained at the beginning of this chapter, lays in an inadequate physical formulation and numerical implementation. In particular the turbulence in the near-wall region cannot be predicted correctly. Since the turbulence variables are the most predominant inputs in the heat-transfer formulation based on the wall function, it becomes evident that even an improved formulation will not be able to achieve a general validation until its inputs are reliable.

10.2.4.1 Sensitivity Analysis of the 3D-CFD-Heat-Transfer calculated with a Wall-Function Model

A sensitivity analysis of the 3D-CFD-heat-transfer calculation with a traditional wall-function model (WF) using as test object a HYDRA-research-engine from Ricardo with four different mesh structures or refinement degrees has been performed (see Figure 10.7).

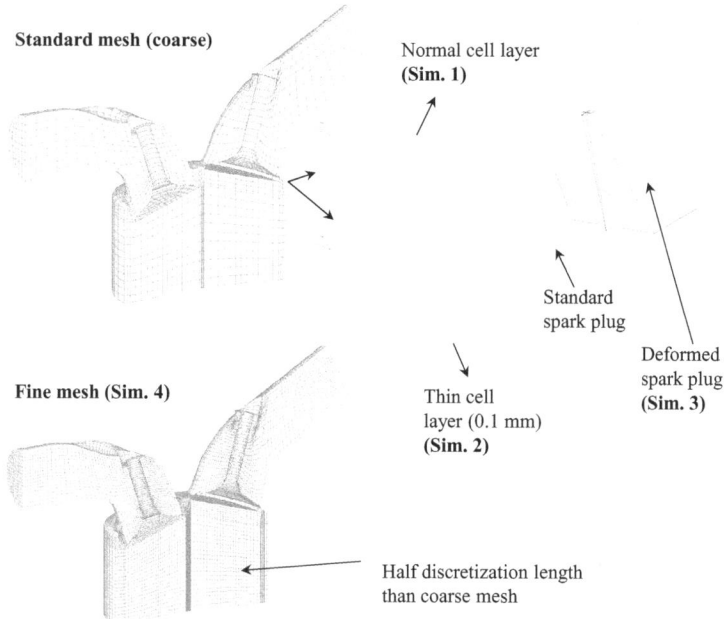

Figure 10.7: *Sensitivity analysis with four different mesh structures or refinement degrees (HYDRA research-engine – 2 valves).*

The simulated engine has a simple combustion chamber shape with a very high mesh quality so that a modification of the mesh can be well targeted and accurate investigations of their effects on the simulation results can be performed. The four meshes of the two-valve research engine differ as follows:

- (Sim. 1 – "Standard") – Standard coarse mesh with ca. 15,000 cells in the combustion chamber.

- (Sim. 2 – "Thin Layer") – Standard coarse mesh with a thin cell layer of 0.1 mm at the cylinder liner.

- (Sim. 3 – "Bad SP") – Standard coarse mesh with a deformed spark plug ("bad" spark plug with locally high aspect ratios and warpage-angles of the cells).

- (Sim. 4 – "Fine") – Fine mesh with half discretization length compared to the standard coarse one (~120,000 cells in the combustion chamber).

The simulated operating condition reported here is 2,000 rpm at WOT with the same boundary conditions. The turbulence is modeled by a $\tilde{k} - \tilde{\varepsilon}$ standard model.

As expected, during the exchange period the average cylinder turbulence TKE $k(\varphi)$, due to high local velocity gradients, shows remarkable differences only with the simulation of the fine mesh (see Figure 10.8). In contrast the profile of $k(\varphi)$ during the working period, in all cases, shows much smaller discrepancies. Since the influence of the turbulence on the heat-transfer and consequently on the engine energy-balance is moderate during the exchange period, the overall variable $k(\varphi)$ can be assumed as a reliable input for heat-transfer modeling.

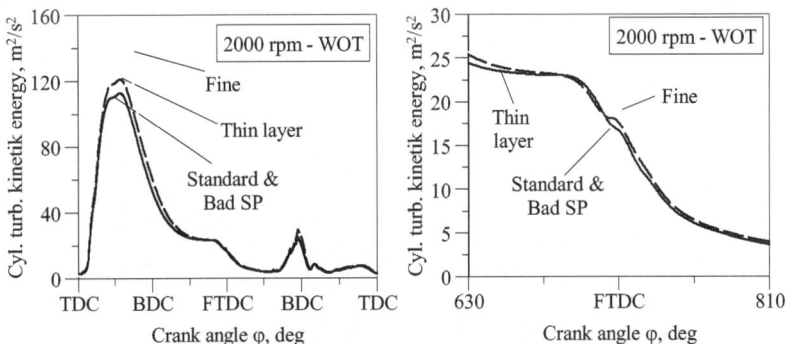

Figure 10.8: *Turbulence TKE in the cylinder during the whole cycle and the working period.*

Looking now at the local values of the turbulence within the cells at the wall where the wall function is applied (boundary layer cells), the differences become more considerable (see Figures 10.9 and 10.10). Focusing on the working period, the influences of dimension, orientation and structure quality of the wall-cells on the turbulence becomes evident. In a moving complex geometry there is no possibility to control any cell and eventually isolate "perturbations" on the local turbulence calculation. Therefore the simulation of the local heat-transfer with a wall-function approach is very often practically "out of control" since one of its most important input variables is not reliable.

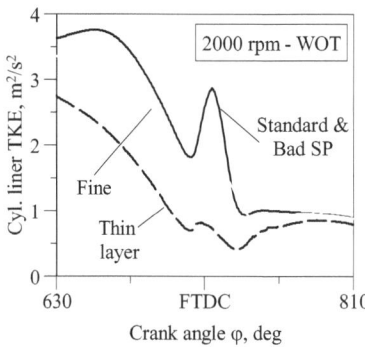

Figure 10.9: *Average TKE at the cylinder-liner wall-cells.*

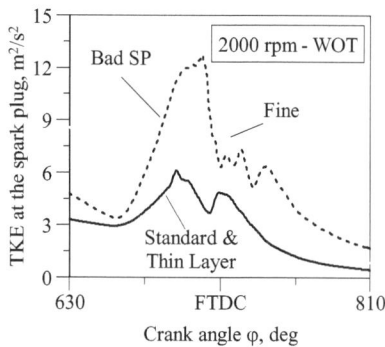

Figure 10.10: *Average TKE at the spark-plug wall-cells.*

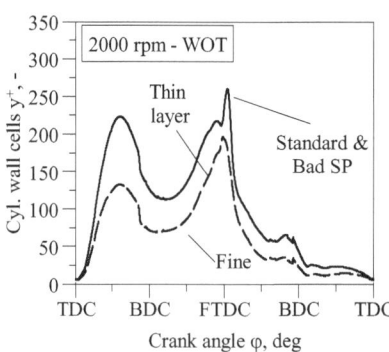

Figure 10.11: *Average y^+-value of the cylinder wall-cells.*

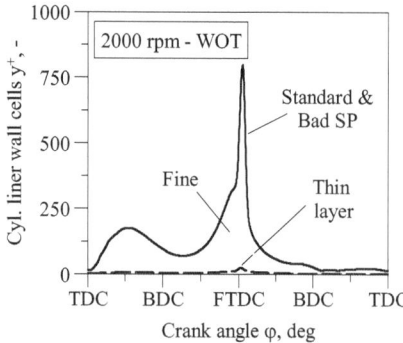

Figure 10.12: *Average y^+-value of the cylinder-liner wall-cells.*

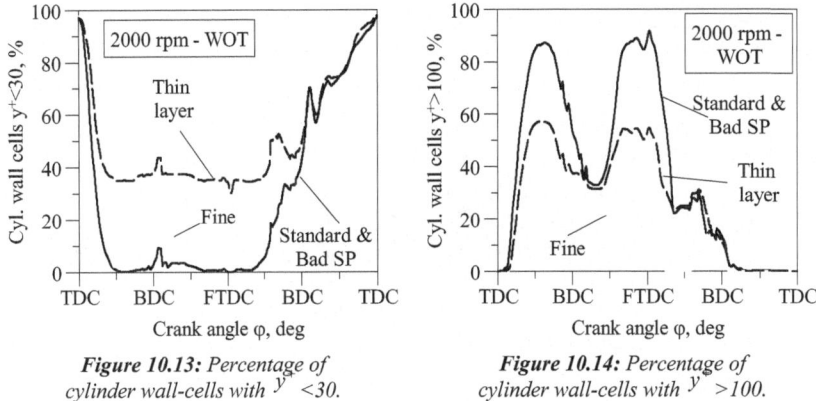

Figure 10.13: *Percentage of cylinder wall-cells with y^+ <30.*

Figure 10.14: *Percentage of cylinder wall-cells with y^+ >100.*

The optimal dimension for the wall-function approach is when the dimensionless wall-distance y^+ assumes values between 30 and 100 units (see Chapter 10.2.1). Here the analysis of y^+ for the four simulations points out the differences to this assumption and the mesh sensitivity on the results (see Figures 10.11-10.14). The y^+-value depends not only on the local mesh structure but also on the flow motion whose transient state varies locally and temporally with high gradients. Nevertheless the flow motion in general depends on the engine operating condition. Therefore if it were possible to tune the mesh structure and the mesh motion for fulfilling the y^+-conditions at any location and time step, this process would be definitely compromised at a different operating condition. For the above mentioned reasons the wall function approach in internal combustion engines cannot rely on optimal y^+-conditions.

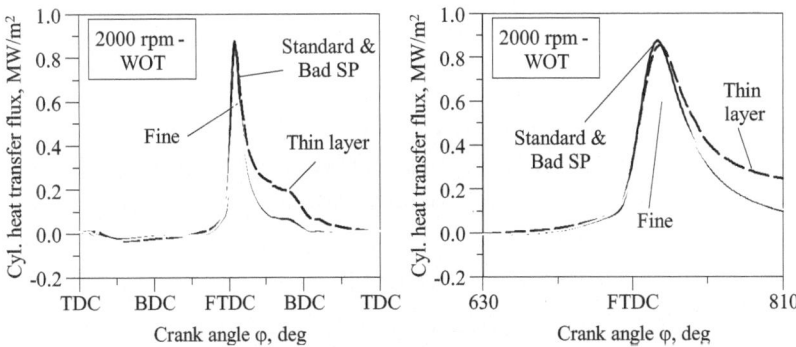

Figure 10.15: *Average heat-transfer flux through the cylinder walls during the whole cycle and the working period.*

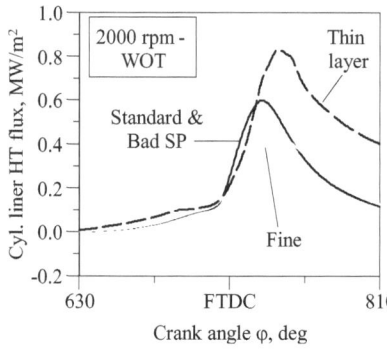

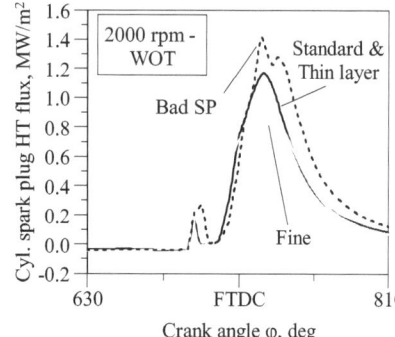

Figure 10.16: *Average heat-transfer flux through the cylinder-liner walls.*

Figure 10.17: *Average heat-transfer flux through the spark plug walls.*

Concluding the analysis of the heat-transfer flux using the wall-function approach (see Figures 10.11-10.12) reports the extreme high sensitivity of the heat-transfer flux with the mesh. The results of the different simulations show variations of about 500% in the prediction of the local heat-flux (in certain cases even more) and up to 100% in the prediction of the overall heat-transfer energy Q_W (J/cycle). The resulting predictability in the calculation of the heat-transfer and the energy balance is obviously not acceptable for a reliable engine analysis.

10.2.5 *QuickSim's* Approach: A new Phenomenological Heat-Transfer Model in the 3D-CFD-Simulation

In this work a three-dimensional phenomenological heat-transfer coefficient $\alpha(\vec{x}, \varphi)$, which is based on the before described Re-Nu-correlations (see Chapters 10.2.3.3 and 10.2.3.4), is proposed for both the working cycle $\alpha_1(\vec{x}, \varphi)$ and the charge changing period $\alpha_2(\vec{x}, \varphi)$. This model is implemented locally, like the wall function, in only a single wall-cells layer, so that computing overheads are not caused. The classic wall function is still used for assuring the no-slip velocity condition and the temperature condition at the wall.

The modeling of the heat-transfer coefficients chosen for the 3D-CFD-simulation, similarly to the real working-process analysis, is based on the following correlations [21,22]:

- Bargende's correlation during the compression and expansion stroke (working cycle).

- Hohenberg's correlation during the intake and exhaust stroke (gas exchange period).

10.2.5.1 The Heat-Transfer during the Working Cycle

The effects of the heat transfer on the working cycle behavior are remarkable because, e.g. the ratio of the heat losses to the total heat transfer usually amounts to more than 80% and it is still more than 45% during the combustion period only.

The aim of this work is to develop a three-dimensional heat transfer model $\overline{\alpha}_1(\vec{x}, \varphi)$, which, first of all, ensures a high accuracy in predicting the overall heat transfer rate $(dQ_W/d\varphi)_{CFD}$ at every time step. This condition is a prerequisite for a thorough prediction of the engine energy balance. In this manner, the overall heat transfer rate calculated from the real working-process analysis (WP) $(dQ_W/d\varphi)_{WP}$ is taken as the representative value.

Target

$$\left(\frac{dQ_W}{d\varphi}\right)_{CFD} = \left(\frac{dQ_W}{d\varphi}\right)_{WP} \tag{10.18}$$

with

$$\left(\frac{dQ_W}{d\varphi}\right)_{CFD} = \sum_j \alpha_1(\vec{x}_j, \varphi) \cdot \left[T_C(\vec{x}_j, \varphi) - T_W(\vec{x}_j)\right] \cdot A_j \cdot \frac{dt}{d\varphi} \tag{10.19}$$

where $T_C(\vec{x}_j, \varphi)$ is the temperature at the center node of a cell j adjacent to the wall with the surface A_j and a constant temperature $T_W(\vec{x}_j)$ over the time.

Starting from the known formulation of Bargende:

$$\alpha_1(\vec{x}_j, \varphi) = f\left(a, V^{-0.073}, p^{0.78}, \overline{T}^{-0.477}, w^{0.78}, \Delta\right) \tag{10.20}$$

where a is a corrector factor for the local implementation and having the detailed results of the flow field in the 3D-CFD-simulation at disposal, it is now necessary to find out how to set appropriately the variables for each wall-cell j in this formulation.

This is a critical choice which requires not only to identify the phenomenon-relevant value for each variable but also to avoid unreliable variables (e.g. local values of k and ε of wall-cells – see Chapter 10.2.4.1). Many trial-and-error adjustments in setting the variables in $\alpha_1(\vec{x}_j, \varphi)$ have been tested in order to reach the best compromise between overall and local heat transfer rate prediction, high accuracy on different engines, high accuracy at different engine operating conditions and low influence of the mesh structure on the results.

Model Development Step

The development tests of $\alpha_1(\vec{x}_j, \varphi)$ have been performed by using both the evaluation tool and the real working-process analysis program (WP) implemented in *QuickSim* via user-subroutines (see Chapter 7). This becomes an "internal 3D-CFD-WP coupling" (see Figure 10.18) as the next feasible step of the result comparison shown in Figure 10.5. This approach ensures not only the same "thermodynamic conditions" between 3D-CFD and WP, but also provides, at every time step, the necessary average values of the variables used in the real working-process analysis to the three-dimensional simulation.

3D-CFD-Simulation

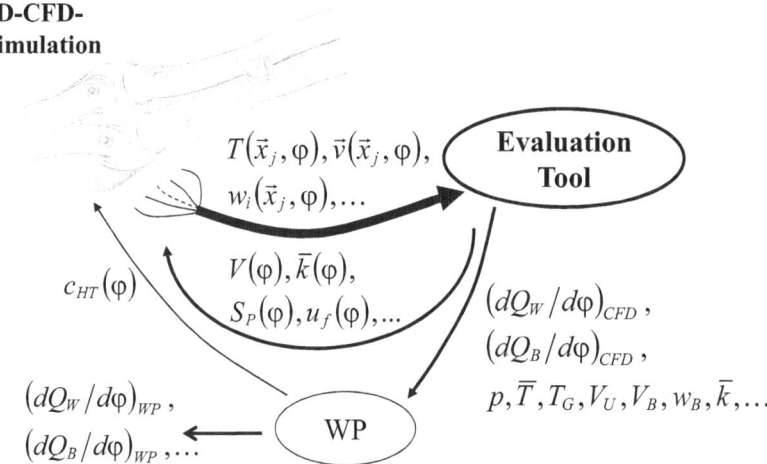

Figure 10.18: *The decisive step: the "internal coupling" between a 3D-CFD-code and the real working-process analysis.*

Choosing a variant of $\alpha_1(\vec{x}_j, \varphi)$, which is mostly a combination of global (averaged overall variables) and local variables, respectively, the heat-transfer rate of the CFD-simulation $(dQ_W/d\varphi)_{CFD}$ has been calculated and compared to that of the working-process analysis $(dQ_W/d\varphi)_{WP}$. If they differed, at every time step, an overall correction variable $c_{HT}(\varphi)$ (see Eq. 10.21) is determined and set in the 3D-CFD-simulation (see Eq. 10.22), in order to correct the wall heat flux (see Eq. 10.23). In this manner at least, during the model developing phase, a representative overall heat transfer rate has been imposed as a boundary condition.

$$c_{HT}(\varphi) = \left(\frac{dQ_W}{d\varphi}\right)_{CFD} \Big/ \left(\frac{dQ_W}{d\varphi}\right)_{WP} \tag{10.21}$$

With the calculation of $c_{HT}(\varphi)$ the heat-transfer coefficient has to be corrected:

$$\alpha_1(\bar{x}_j, \varphi)_{corr} = c_{HT}(\varphi) \cdot \alpha_1(\bar{x}_j, \varphi) \tag{10.22}$$

so that:

$$\left(\frac{dQ_W}{d\varphi}\right)_{CFD,corr} = c_{HT}(\varphi) \cdot \left(\frac{dQ_W}{d\varphi}\right)_{CFD} = \left(\frac{dQ_W}{d\varphi}\right)_{WP}. \tag{10.23}$$

The phenomenological model was not validated until the heat-transfer correction variable $c_{HT}(\varphi)$ became approximately a constant c_{HT}, so that this value can be included in the constant a of Eq.10.20, leading to:

$$\alpha_1(\bar{x}_j, \varphi) = f\left(\bar{a}, V^{-0.073}, p^{0.78}, \bar{T}^{-0.477}, w^{0.78}, \Delta\right) \tag{10.24}$$

where:

$$\bar{a} = a \cdot c_{HT} \tag{10.25}$$

In addition to the above mentioned approach that uses the validated model with a constant value of c_{HT}, the user of *QuickSim* has also the possibility to run a simulation with an automatic adjustment of $c_{HT}(\varphi)$. In this case the profile of the calculated 3D-CFD-heat-transfer (actual value) will always equal that one of the real working-process analysis (set value).

Final Formulation

In this section the numerical implementation procedure of each variable in the phenomenological model is described for the final version of $\alpha_1(\bar{x}_j, \varphi)$:

Volume

The instantaneous cylinder volume $V(\varphi)$ is taken as the characteristic geometrical entity in which the heat-transfer process takes place.

Pressure

The pressure in the combustion chamber of an internal combustion engine can be assumed as uniform. The local value of the pressure in the wall-cell $p(\varphi)$ is taken as the relevant value.

Temperature

In order to calculate the local relevant gas temperature $\bar{T}(\vec{x},\varphi)$ the prediction of the local gas temperature outside the thermal boundary layer $T_G'(\vec{x},\varphi)$ is necessary:

$$\bar{T}(\vec{x}_j,\varphi) = \frac{T_G'(\vec{x},\varphi) + T_W(\vec{x}_j)}{2} \tag{10.26}$$

The determination of $T_G'(\vec{x},\varphi)$ requires a high accuracy because in many situations it differs completely from the temperature $T_C(\vec{x}_j,\varphi)$ of the wall-cell which is provided by the CFD-code (see Figure 10.19). As mentioned in Chapter 10.2.1, the central node P of the wall-cell should not be located too close to the wall (case 2 of Figure 10.19) in order to fulfill the y^+-condition. But especially near TDC due to the high compression degree of the cell layer, it is not possible to satisfy this condition at many locations.

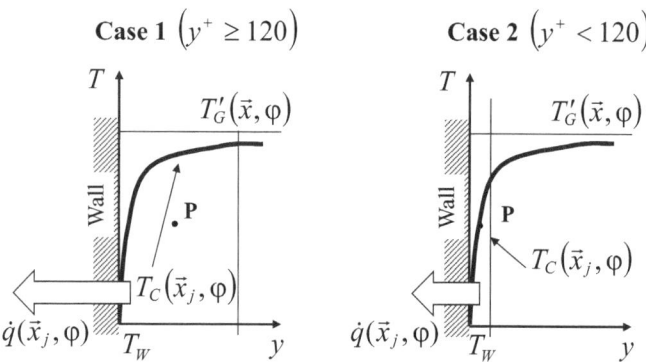

Figure 10.19: Effect of cell thickness on the determination of the cell temperature.

For that goal, several numerical investigations on engine test meshes have been carried out for deducing the relations among the thickness of the thermal boundary layer, $T_G'(\vec{x},\varphi)$ and $T_C(\vec{x}_j,\varphi)$. Nevertheless the temperature $T_G'(\vec{x},\varphi)$ is needed for the correct determination of the local real temperature difference $\Delta T' = [T_G'(\vec{x},\varphi) - T_W(\vec{x}_j)]$ between wall and gas-side (see Eq. 10.19), which differs from the numerical assumed $\Delta T = [T_C(\vec{x}_j,\varphi) - T_W(\vec{x}_j)]$ in the CFD-code. It has been found out that the following relations allow both high solution accuracy and very low influence of the mesh structure on the results:

- **Case 1** - $y^+ \geq 120$

$$T_G'(\vec{x},\varphi) \cong T_C(\vec{x}_j,\varphi) \tag{10.27}$$

and

$$\dot{q}_W = \alpha_1(\vec{x}_j, \varphi) \cdot \left[T_C(\vec{x}_j, \varphi) - T_W(\vec{x}_j) \right]. \tag{10.28}$$

- **Case 2 -** $y^+ < 120$

In this case, the temperature $T'_G(\vec{x}, \varphi)$ must be extrapolated and a correction factor $C_T(\vec{x}_j, \varphi)$ for the heat-transfer flux calculation has to be included:

$$T'_G(\vec{x}, \varphi) = f\left(T_W(\vec{x}_j), T_C(\vec{x}_j, \varphi), y^+, \ldots \right) \tag{10.29}$$

and

$$\dot{q}_W = \alpha_1(\vec{x}_j, \varphi) \cdot C_T(\vec{x}_j, \varphi) \cdot \left[T_C(\vec{x}_j, \varphi) - T_W(\vec{x}_j) \right] \tag{10.30}$$

where

$$C_T(\vec{x}_j, \varphi) = \frac{\left[T'_G(\vec{x}, \varphi) - T_W(\vec{x}_j) \right]}{\left[T_C(\vec{x}_j, \varphi) - T_W(\vec{x}_j) \right]} \tag{10.31}$$

so that:

$$\dot{q}_W = \alpha_1(\vec{x}_j, \varphi) \cdot \left[T'_G(\vec{x}, \varphi) - T_W(\vec{x}_j) \right]. \tag{10.32}$$

A simple but satisfying relation, which enables to extrapolate the required temperature $T'_G(\vec{x}, \varphi)$ has been carried out for this goal. The following relation is based on the assumption of a logarithmic temperature profile within the boundary layer.

$$T'_G(\vec{x}, \varphi) = T_W + (T_C - T_W) \cdot \frac{\ln 120}{\ln(y^+ + 1)} \tag{10.33}$$

Velocity and combustion term

Many attempts to take the local values of both turbulence and velocity of the wall-cell as the representative turbulent kinetic energy and fluid velocity, respectively, have not achieved a general validation. As explained in Chapter 10.2.4.1, especially the sensitivity of the local turbulence of the heat-flux calculation makes this variable, up to now, unreliable for the heat-transfer calculation. Therefore the global average turbulence in the cylinder $k(\varphi)$ has been chosen for the final formulation:

$$\bar{k}(\varphi) = k(\varphi). \tag{10.34}$$

Similarly the instantaneous piston speed $S_P(\varphi)$ remains in the formulation of the velocity term.

During the development tests of $\alpha_1(\bar{x}_j, \varphi)$, many numerical investigations have been carried out in order to take appropriately into account the effect of the combustion process on the engine heat-transfer. It has been found out and widely proved that the spatially-averaged flame propagation speed $u_f(\varphi)$ (absolute flame speed) is a feasible variable for considering the combustion process (see e.g. Figure 10.20 – MB M102-E23 – two-valve engine). The flame speed $u_f(\varphi)$ has then been embodied in the velocity term $w(\varphi)$ (for the burned zone only).

- Unburned zone

$$w(\varphi) = \frac{1}{2}\sqrt{\frac{8}{3}\bar{k}(\varphi) + S_P(\varphi)^2} \tag{10.35}$$

- Burned zone

$$w(\varphi) = \frac{1}{2}\sqrt{\frac{8}{3}\bar{k}(\varphi) + S_P(\varphi)^2 + u_f(\varphi)^2} \tag{10.36}$$

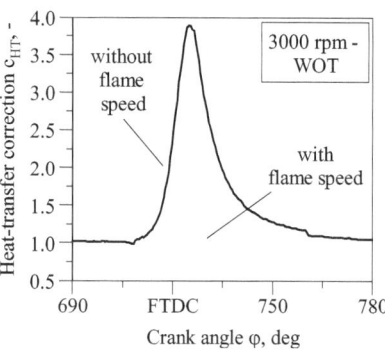

Figure 10.20: *Variation of the corrector factor $c_{HT}(\varphi)$ with and without considering the flame speed $u_f(\varphi)$ in the velocity term $w(\varphi)$.*

Constant $\bar{a}$

The constant $\bar{a}$, during the validation step, did not require any adjustment $(c_{HT} = 1)$:

$$\bar{a} = a = 253.5 \tag{10.37}$$

Conclusion

Concluding, the three-dimensional phenomenological heat transfer $\alpha_1(\bar{x}_j, \varphi)$ is then given by:

$$\alpha_1(\bar{x}, \varphi) = 253.5 \cdot V(\varphi)^{-0.073} \cdot p(\bar{x}_j, \varphi)^{0.78} \cdot \overline{T}(\bar{x}_j, \varphi)^{-0.477} \cdot w(\varphi)^{0.78}. \quad (10.38)$$

Additionally, in the heat flux calculation, a correction factor $C_T(\bar{x}_j, \varphi)$ has to be used for wall-cells which do not satisfy the condition $y^+ \geq 120$ (see Eqs. 10.27-10.33).

10.2.5.2 The Heat-Transfer during the Charge Changing Period

The same procedure as undertaken for the working cycle has also been followed for the gas exchange period. Briefly described, the three-dimensional phenomenological modeling of the local heat transfer coefficient $\alpha_2(\bar{x}_j, \varphi)$ at the wall cell j is here based on a local application of Hohenberg's correlation (see Chapter 10.2.3.3). This model is implemented in a single wall-cells layer.

$$\alpha_2(\bar{x}_j, \varphi) = 130 \cdot V(\varphi)^{-0.06} \cdot p(\bar{x}_j, \varphi)^{0.8} \cdot T_G'(\bar{x}, \varphi)^{-0.4} \cdot (\overline{S}_P + 1.4)^{0.8} \quad (10.39)$$

The validity of the above presented relations for the determination of $T_G'(\bar{x}, \varphi)$ and $C_T(\bar{x}_j, \varphi)$ is still maintained (see Eqs. 10.27-10.33).

10.3 Results

Here are shown some results of the three 3D-CFD-simulations using the proposed phenomenological heat-transfer model (3D-Ph.). The simulations refer to different operating conditions of a mass-production SI-engine with two valves (MB M102-E23 – Engine specifications in Case 1 of Table 10.1). The simulated pressure profiles have been compared with measurements at the test bench (Exp.) while the predicted 3D-CFD-heat-transfer rates have been compared with the results of the real working-process analysis (WP).

The three simulated operation conditions reported here are:

- 3,000 rpm – WOT – imep = 10.8 bar

- 4,000 rpm – WOT – imep = 11.3 bar

- 1,500 rpm – Low Load – imep = 3.4 bar

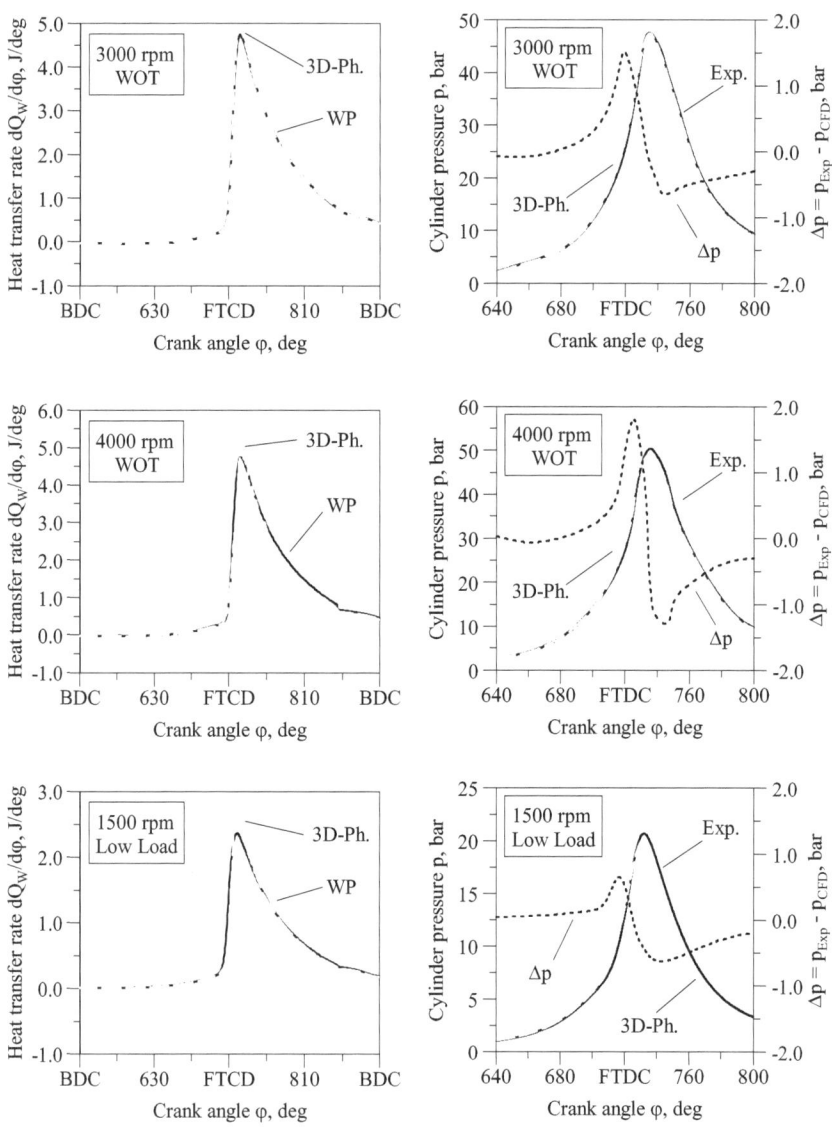

Figure 10.21: *Heat-transfer rate.*
Comparison 3D-Ph. vs. WP.

Figure 10.22: *Cylinder pressure.*
Comparison 3D-Ph. vs. Exp.

The results of the proposed phenomenological heat-transfer model (3D-Ph.) show high accuracy in predicting the engine heat-transfer with no associated computing overheads. Also at low-load operating condition the predicted heat-transfer rate remains plausible and in accordance with the real working-process analysis.

10.4 Influence of 3D-CFD-Heat-Transfer-Models on the Engine Energy-Balance

An accurate prediction of the engine heat-transfer is a prerequisite for a reliable simulation of many other thermodynamic processes which take place in the combustion chamber. The mentioned effects of the heat-transfer on the results of an engine simulation, in particular on the energy balance, are briefly reported below by comparing the results of two simulations, which differ only from the implemented heat-transfer model: a 3D-CFD heat-transfer model based on the wall function approach (3D-WF) and the proposed phenomenological 3D-CFD heat-transfer model (3D-Ph.).

Table 10.2: Comparison of the engine energy-balance.
(3D-CFD vs. Experiments) – 3,000 rpm – WOT (SI-engine: MB M102-E23 with 2 valves).

	3D-WF	3D-Ph.	Experiments
η_V (vol. efficiency), -	0.79	0.785	0.775
Imep, bar	12	10.9	10.8
p_{max}, bar	50.5	47.6	47.3
Isfc, g/kWh	220	242	245
Heat release Q_B, J/cycle	1684	1588	-
Heat transfer Q_W, J/cycle	63	335	-
Work, J/cycle	689	626	620
Intake and exhaust enthalpy, J/cycle	932	627	-

Switching from the phenomenological approach (3D-Ph.) to the wall-function (3D-WF), mainly an implausible relevant "shifting" of the energy balance from the wall heat-transfer to the exhaust enthalpy occurs (see Figure 10.23). Here, as reported in Figure 10.25, a remarkable increasing of the temperature in the cylinder is evident (up to 200 K in the expansion stroke).

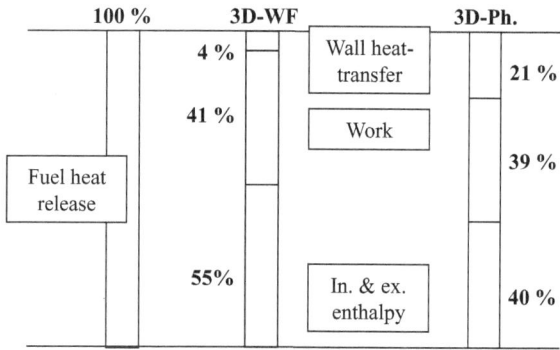

Figure 10.23: *Engine energy-balance. 3D-CFD-heat-transfer model comparison: phenomenological (3D-Ph.) vs. wall-function (WF).*

Since the heat-transfer calculation has a negligible influence on the volumetric efficiency η_V (see Table 10.2) the temperature arising within the combustion chamber is related to an increase of the cylinder pressure (see Figure 10.24) which consequently leads to an overestimation of the produced work. However, between the two heat-transfer models the variation of work is much smaller than the variation of exhaust enthalpy.

The differences of temperature in the combustion chamber during combustion are evident in both the unburned and the burned zone, respectively (see Figures 10.25 and 10.26). In the unburned zone the wrong prediction of temperature leads meanly to inaccurate analyses of:

- Knock sensitivity.

- Flame propagation speed.

- Flame quenching distance (UHC-emissions, etc.).

Similarly the wrong prediction of the temperature in the burned zone causes mainly inaccurate analyses of:

- NO_x-emissions.

- Thermodynamic properties of burned gas (e.g. dissociation effects).

- Combustion efficiency.

- Temperature of EGR.

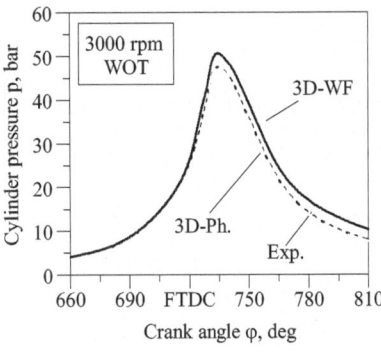

Figure 10.24: Cylinder pressure.

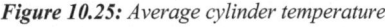

Figure 10.25: Average cylinder temperature.

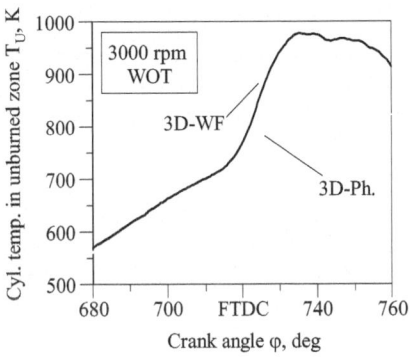

Figure 10.26: Average temperature in the cylinder unburned-zone T_U.

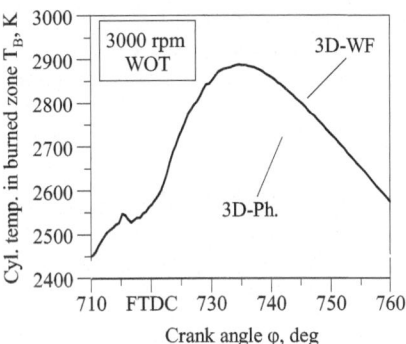

Figure 10.27: Average temperature in the cylinder burned-zone T_B.

Concluding, the improvement in calculating the engine heat-transfer, obtained by the proposed phenomenological heat-transfer model, permits to simulate the engine energy balance with high accuracy, so that the thermodynamic conditions for other thermodynamic models (e.g. combustion-model, NO_x-model, knock-model) can be set more correctly.

11

A Way towards Virtual Engine Development

In this chapter a recent application of *QuickSim* will be presented (year: 2009). The focus here is mainly on the analysis of different approaches for supporting the engine development process: from a "simple" three-dimensional visualization and understanding of the occurring phenomena within the combustion chamber or the airbox up to a reliable virtual engine development. The discussion of these approaches aims to underline the advantages, the drawbacks, the reliability and the predictability of each of them. The investigated engine, as test carrier, is a turbocharged Compressed Natural Gas engine that because of its peculiarities represents a very innovative and promising, but also complex solution in the future engine development.

11.1 Introduction

At the 2009 edition of the 24-hour endurance race on the Nürburgring the Volkswagen Motorsport GmbH, in addition to three Scirocco GT24 vehicles powered by gasoline engines, used two Scirocco vehicles powered by innovative CNG engines for the first time. This wanted to proof, that also different concepts are able to deliver the required efficiency, dynamic and reliability for a successful participation in Motorsports [64,87].

11.2 The Hardware: a turbocharged CNG Race-Engine

For this endurance test Volkswagen developed a CNG engine based on the two-litre gasoline TSI engine (four cylinders, 16 valves) of the previous race edition (2008). The engineers primarily focused on the intake system of the unit equipped with a turbocharger as well as on the balance between maximum exhaust gas temperature and full usage of the turbocharger potentiality in order to optimize both drivability and power output.

Table 11.1: Vehicle specifications: VW Scirocco GT 24 (GDI-engine) vs. GT 24-CNG.

Specifications:	GT 24	GT 24 CNG
Curb weight, kg	1,000	1,130
Tank capacity	100 liters (gasoline)	44 kg (natural gas)
Engine displacement, cm^3	1,998	1,998
Max. boost pressure, mbar	2,400	2,400
Ø Air-restrictor, mm	38	38
Fuel injection	Direct injection	Manifold injection
Max. torque, Nm	370 at 4,500 rpm	330 at 3,500 rpm
Max. power, kW	232	207

In Table 11.1 the specifications of the vehicles with the two different engine concepts are reported. Due to the lower mass flow of the CNG-fuel-injector, the CNG-engine is equipped with a manifold fuel injection system (two injectors per cylinder) instead of direct injection. The maximal boost pressure of the turbo-charging system is limited by regulation to 2,400 mbar in combination with an air-restrictor (Ø 38mm).

In the CNG version the limitation of the boost pressure causes a reduction of the maximal power of about 10%. This performance reduction is the consequence of the loss of charge due to the gas injection in the intake manifold. Because of both the high volumetric displacement of natural gases and the absence of heat of vaporization the intake air flow at the same boost pressure is lower in comparison to the gasoline version. Reclusively, the higher hydrogen content of methane which represents the main component of NG leads to a CO_2 reduction during combustion of about 20% compared to gasoline [33,88-94]. Thanks to the laminar burning speed of methane which is approximately maximal for a stoichiometric mixture, instead of approximately λ=0.9 for gasoline, CNG engines do not require a mixture enrichment at WOT operating conditions, so that the fuel consumption decreases [95]. In addition the very high

RON-equivalent of natural gas allows a further efficiency increasing by using a higher compression ratio [96,97]. Conclusively the CO_2 reduction (g/km) of the CNG-version is from ca. 30% using natural gas up to 80% using bio-gas.

Figure 11.1: *The turbo-charged CNG-engine at the test bench.*

Figure 11.1 shows the engine at the test bench. Oil and water temperatures are conditioned. In addition to the common measurement devices the engine is equipped with a full indicating system for high and low pressure analysis in each cylinder. In addition there are pressure sensors for the airbox and the turbine. Low pressure sensors in the intake manifold are between the airbox and the fuel injectors and in the exhaust manifold at about 50 mm from the cylinder head.

11.3 Setting of the 3D-CFD-Simulation

The following 3D-CFD-simulations with the fast response code *QuickSim* have been performed using four different extensions of the domain of the 3D-CFD-mesh:

- One cylinder with parts of intake and exhaust channel (the fuel injectors are outside the mesh).

- One cylinder with parts of intake and exhaust channel up to the locations of the low pressure sensors (the fuel injectors are included in the mesh).

- One cylinder in combination with the whole airbox.

- Full engine from the airbox up to the turbine.

The four meshes have the same structure in the common parts and have been generated using the half-automatic mesh-generation-program of *QuickSim* for the moving parts and common mesh generation programs (in this case ProAm) for all other engine parts. The cells in the combustion chamber are exclusively hexahedral and starting from the static mesh, the cylinder motion is automatically provided by *QuickSim* using cell compression (no layer delay) for the piston movement.

Table 11.2: 3D-CFD-Mesh specifications and calculation data.

Specifications:	Cylinder w/o Injectors	Cylinder with Injectors	Cylinder with Airbox	Full Engine
Mesh cell number, -	69,000	75,000	190,000	450,000
Average discr. length in the cylinder l_D, mm	1.8	1.8	1.8	1.8
Average discr. length in the airbox l_D, mm	-	-	2.5	2.5
No. of fuel parcels (cyl./cycle), -	-	110,000	110,000	110,000
CPU-time, h/cycle	3	3.5	18	24
No. of cycles to convergence at WOT, -	3	3	7	5
Memory of executable File, MB	60	90	200	550
Disk-space, GB/cycle	2	10	20	60

In Table 11.2 the specifications of the four meshes are reported. Here it can be seen that also in case of a full-engine simulation with only one processor the computational time still remains acceptable within the timetable of engine development. As mentioned in Chapter 7.1.5.1, during a project, the target here is the simultaneous simulation of different operating conditions or engine designs instead of a complex multi-processor run of only one simulation.

Table 11.3: Specifications of the simulated operating conditions.

Specifications:	Sim. 1	Sim. 2	Sim. 3
rpm, -	5,500	6,000	6,500
Load, -	WOT	WOT	WOT
Injected fuel, kg/h	100 %	113.8 %	111.3 %
Air consumption, kg/h	100 %	108.1 %	105.6 %
Torque, Nm	100 %	103.4 %	83.0 %
Power, KW	100 %	105.3 %	97.7 %

Table 11.3 reports a selection of the numerous simulated operating conditions that will be presented and discussed in this work. Due to secrecy reasons these simulations refer to the basis engine used during the race edition 2009 (normalized values with 100% value for simulation No. 1), i.e. further improvements undertaken during the last months are not reported. All the simulated operating conditions are at full load which is the most relevant running condition of a race engine.

11.3.1 Initial Conditions and Properties of the Working Fluid

The initial conditions within the 3D-CFD-domain, as introduced in Chapter 7.2, are set dividing the mesh into certain macro-regions: one for each cylinder, the intake system and the exhaust system. In each macro-region the temperature, pressure and mixture composition are almost constant. An exception, e.g., is the possibility to set a different mass fraction of fuel (fuel pre-load) in the intake channels within the cylinder head. This allows a faster "running-in phase" of engine simulations with a manifold fuel injection where parts of the fuel in the cylinder is injected in the previous cycles. In absence of reliable information about the fluid motion in all parts of the discretized engine no fluid velocity is set as initial condition. In this project the initial conditions have been provided by the real working-process analysis for the cylinders and 1D-CFD-programs for the intake and exhaust system, respectively.

The modeling of the properties of the working fluid have been performed according to the description in Chapter 8. The detailed properties of exhaust gases are reported in Appendix B.

11.3.2 Boundary Conditions

The boundary conditions at the manifolds of the 3D-CFD-mesh (switching region between 1D-CFD and 3D-CFD domain – see Chapter 7.2), depending on the extension of the domain, are pressure traces from low-pressure sensors or in case of different locations they are calculated using 1D-CFD-simulation tools (rarely, if a well calibrated model is missing they have to be estimated). Until a reliable coupling (conjunct simulation) between *QuickSim* or in generally 3D- and 1D-CFD-codes is not available, in all cases the setting of boundary conditions is performed according to the following points:

Pressure vs. Mass Flow

Pressure curves p instead of mass flows have been imposed. This is always the preferable way because it allows increasing the predictability of the program in case of simulations of different virtual variants; otherwise, e.g., one of the most relevant variables like the intake mass flow is imposed a priori. The velocity at the boundary conditions is then calculated by an internal 1D-CFD-model implemented in *QuickSim* that allows a reliable propagation of the pressure waves through the whole manifolds even when only a part is included in the 3D-CFD-domain.

Temperature and Fluid Composition

The setting of the temperature and the fluid composition at the manifold boundary conditions is a crucial process (see Figure 11.2). If the mass flow is directed inwards the 3D-CFD-mesh, the pressure, velocity, temperature and the composition of the inflowing fluid (w_{Air}, w_F and w_{EGR}) are imposed by the setting of the boundary-conditions. If the mass flow is directed outwards the 3D-CFD-mesh, only the pressure and velocity are imposed by the boundary-conditions. Other variables, like the temperature and the composition of the outflowing fluid ("backflow") are set equal to the mass-averaged values of these variables in the 3D-boundary cells. If incongruences between the in- and backflow take place remarkable errors (e.g. in the mass conservation of the species and in particular fuel and air) may occur, which irremediably compromise the reliability of the simulation [88].

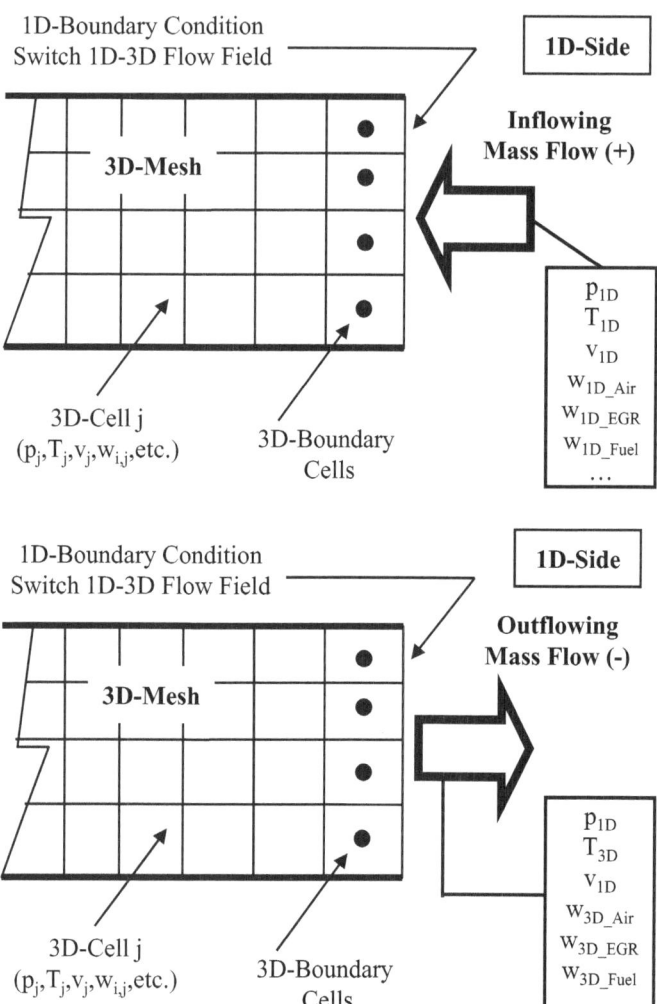

Figure 11.2: *Setting of the manifold boundary conditions.*

The setting of the temperature T can be performed using data from 1D-CFD-simulations or, preferably like in this project, they can be directly calculated within *QuickSim* using the following equation:

$$T = T_0 \cdot \left(\frac{p}{p_0} \right)^{\frac{n-1}{n}} . \tag{11.1}$$

Here a reference temperature T_0 for a reference pressure p_0 (usually 1 bar) and a polytropic coefficient n (as function of the fluid composition) are set according to a one-time calibration process of the intake air mass.

The setting of the scalars for the description of the composition of the working fluid, as introduced before, is even more complex when mixing processes through the boundary conditions take place. Especially the fuel mass-balance at the intake boundary conditions, depending on the location of the fuel injectors and in case of intense backflow streams (see Chapter 11.5), requires often an adjustment of the scalars in order to ensure the correct fuel and EGR mass in the cylinder at IVC. Concluding, as explained in Chapter 7.2, the best locations for the boundary conditions are where the manifolds are straight with a constant cross section where mixing processes between air, fuel or exhaust gas do not take place.

Fuel Injection

The boundary conditions at the nozzle of the CNG-fuel-injectors are no more than estimations. Only the mass flow can be derived from both the injector specifications and the injection duration. In Chapter 11.4 the way how to reproduce the fuel injection is reported.

11.4 CNG-Injector Model

The modeling of the behavior of the injector in order to determine all the required variables for the 3D-CFD-simulation (fuel mass flow, injection velocity, density, temperature, nozzle jet angles, etc.) is a very complex task. Because of both the extremely high complexity and the exorbitant CPU-time required for a 3D-CFD-simulation which aims to extend the 3D-domain to the internal flow within the injector, in this case as usual, it is convenient to set the properties of the injected methane as boundary conditions at the injector nozzle location.

The determination of the fuel mass flow profile is particularly difficult because, first of all, the injection timing is influenced by the opening delay of the nozzle depending on battery voltage and rail pressure. Another problem is the lack of fluid damping in NG-operation which makes the plunger bouncing at valve opening and closing. Whereas the influence on injection timing of both battery voltage and rail pressure can be taken into account and compensated by the ECU as well, the plunger bouncing still remains a critical point which has great influence on the mixture formation process. In the absence of detailed needle lift data for the CFD-calculation the assumption of a characteristic mass flow was necessary. Therefore, an extrapolated needle lift curve was drawn based on information given by the Bosch GmbH [88]. With this curve, for sub-critical flow, the gas mass flow through the injection valve can be calculated using the equation of Saint-Venant:

$$\frac{dm_{Inj\_F}}{dt} = C_{Inj} \cdot A_{Geom.} \cdot p_{rail} \cdot \sqrt{\frac{1}{R_{rail} \cdot T_{rail}}} \cdot \sqrt{\frac{2 \cdot \kappa}{\kappa - 1}\left[\left(\frac{p_{Intake}}{p_{rail}}\right)^{\frac{2}{\kappa}} - \left(\frac{p_{Intake}}{p_{rail}}\right)^{\frac{\kappa+1}{\kappa}}\right]} \qquad (11.2)$$

Here, the fuel mass flow rate dm_{Inj_F}/dt through the injector is given as a function of the discharge flow coefficient C_{Inj}, the orifice geometric flow area A_{Geom}, the pressure ratio p_{Intake}/p_{rail}, the specific fluid constant R_{rail}, the specific heat ratio $\kappa = c_p/c_v$ and the temperature T_{rail} in the fuel rail. If the pressure ratio p_{Intake}/p_{rail} is less or equal the critical pressure ratio (see Eq. 11.3) the flow of the fuel becomes choked. Under this condition the gas mass flow through the nozzle reaches a maximum (see Eq. 11.4), and it does not depend on the pressure in the intake anymore, only on the condition in the fuel rail.

$$\frac{p_{Intake}}{p_{rail}} \leq \left(\frac{2}{\kappa + 1}\right)^{\frac{\kappa}{\kappa - 1}} \qquad (11.3)$$

$$\left(\frac{dm_{Inj\_F}}{dt}\right)_{max} = C_{Inj} \cdot A_{Geom.} \cdot p_{rail} \cdot \sqrt{\frac{1}{R_{rail} \cdot T_{rail}}} \cdot \sqrt{\frac{2 \cdot \kappa}{\kappa + 1}} \qquad (11.4)$$

In case of a manifold injection (max. p_{Intake} = 2,500 mbar) the pressure ratio is always less than the critical value, i.e. the flow through the injector is always choked. Figure 11.3 shows the needle lift trace and the mass flow rate at 4000 rpm, WOT operating condition. At the end of the valve opening a moderate valve bouncing takes place. It can be seen that these needle oscillations have negligible influence on the mass flow. The reason is the very small change in the effective valve area for higher lift values where the discharge flow coefficient remains practically constant to its maximum (see Figure 11.4).

When the valve closes some more bouncing occurs. This is intensified by the gradient of the discharge flow coefficient at lower valve lift. Thus, depending on both the operating condition (in particular for low load and idle operating conditions) and the number of injections (multiple injection), up to one third of the whole fuel mass can be expected to be injected by the effect of the valve bouncing at closing. Due to the broad variation of the ballistic valve operation in the bouncing mode, fluctuations of the mixture formation can be expected, too.

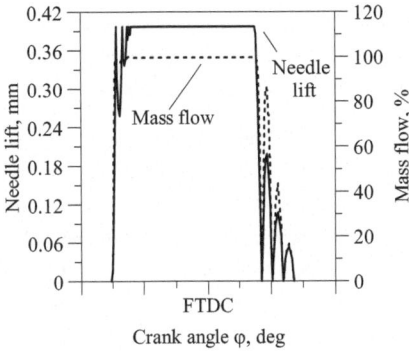

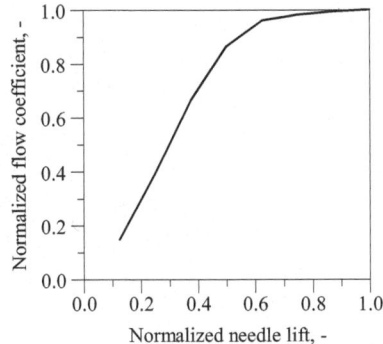

Figure 11.3: *Curves of injector needle lift and gas mass flow.*

Figure 11.4: *Discharge flow coefficient C_{Inj} as a function of normalized needle lift.*

The required injection velocity V_{Inj} can then be calculated from both the injection mass flow and the gas properties at the nozzle. More crucial is the determination of the jet angles α and ε (see Chapter 11.4.2). Whereas the external jet angle α of the injection cone can be measured by experimental investigations in a pressure chamber, the angle ε, particular in case of injectors with a nozzle that generates a hollow cone, cannot be determined experimentally. Due to the absence of experiments, the setting of the angle ε is tuned to fit the measured jet development.

11.4.1 Traditional Gas Injection Modeling

In the traditional procedure the properties of the injected gaseous fuel (mass flow, injection velocity, density, temperature, nozzle jet angles, etc.) are discretized and imposed to 2D-inlet-boundary-conditions (see Figure 11.5), that are represented by the faces of the border cells of the 3D-CFD-mesh at the corresponding location of the injector nozzle. Since the injected fuel has a very high initial velocity (V_{Inj} up to 400 m/s), a perturbation generating extremely high velocity gradients in the cells near the injector nozzle takes place (numeric crucial region). Under these conditions, the calculation runs very often unstable and even a drastic reduction of both time step Δt and under-relaxation factor do not permit to get an adequate convergence of the results.

Another problem is represented by the required discretization of the 3D-CFD-mesh which has to provide the conditions for establishing a realistic shape of the fuel jet. The information of the gaseous fuel concentration starting from the cells where the 2D-inlet-boundary-conditions are applied is passed sequentially to their neighboring cells entirely through the faces of these initial cells by solving the conservation equations for each cell central node (Eulerian formulation) [53,54,56]. Since the Eulerian formulation assumes a homogeneous distribution of all variables within each cell, it is evident that the dimension of the cells, especially in the near nozzle region, drastically influences the shape of the simulated jet and its axial and radial penetration.

Here, a high refinement and adaptation of the mesh with cells much smaller than the injector orifice certainly ensures a better resolution of the flow field, but on the other hand, the exorbitant CPU-time in combination with an increased "numerical instability" of the small cells make this way unjustifiable. Exemplarily in the case of a local cell discretization of 0.1 mm and a fluid velocity equal to the injection velocity V_{Inj}, in order to satisfy the condition imposed by the Courant number (see Eq. 7.1) an exorbitant small computation time $\Delta t = 0.0002$ ms (ca. $\Delta\varphi$ =0.002 deg at 2,000 rpm) is required. In particular, in case of either a multi-hole injector or an injector that generates a hollow jet cone, it is very difficult to find an affordable compromise between the appropriate mesh discretization that permits a good simulation of the jet development and the required efforts.

11.4.2 Gas Injection Modeling in *QuickSim*

A new approach towards an improved gaseous fuel injection has been tested and implemented in *QuickSim*. Here the modeling of the injected fuel in the near nozzle region introduces the concept of gaseous bubbles that numerically are treated as "fictive" gas droplets (see Figure 11.6). This numerical approach is well known in case of liquid injection (spray injection) and in this 3D-CFD-tool it has been adapted to the gaseous fuel injection. In 3D-CFD-codes the charge assumed as a continuous fluid is modeled by a standard Eulerian formulation and the methane bubbles are added to the gas phase using the Lagrange formulation [53,54,56]. Since modeling the millions of individual methane bubbles would lead to a prohibitive computational time, the Lagrange formulation introduces the concept of the parcel which represents a sample of many "fictive" gas droplets (Discrete Droplet Method). The parcels are introduced in the mesh with initial values of position, size D, velocity, temperature and number of "fictive" gas droplets per parcel (Injector Model).

The numerical two-way-coupling of the Eulerian and the Lagrange equations allows the modeling of the exchange processes (mass, momentum and energy) between the charge and the injected methane. Thanks to this approach the number of the information saved in the total amount of the parcels is very high and independent on the discretization degree of the cells in the near nozzle region. The increased CPU-time required for the "fictive" gas droplets remains in an acceptable range. The droplets are introduced in the 3D-CFD-mesh randomly in a space outside the injector orifice at a distance between 0.1 and 2 mm. This space is delimited by a hollow cone generated by the angles α and ε (see Figure 11.6). Also here, a multi-hole injector can easily be modeled.

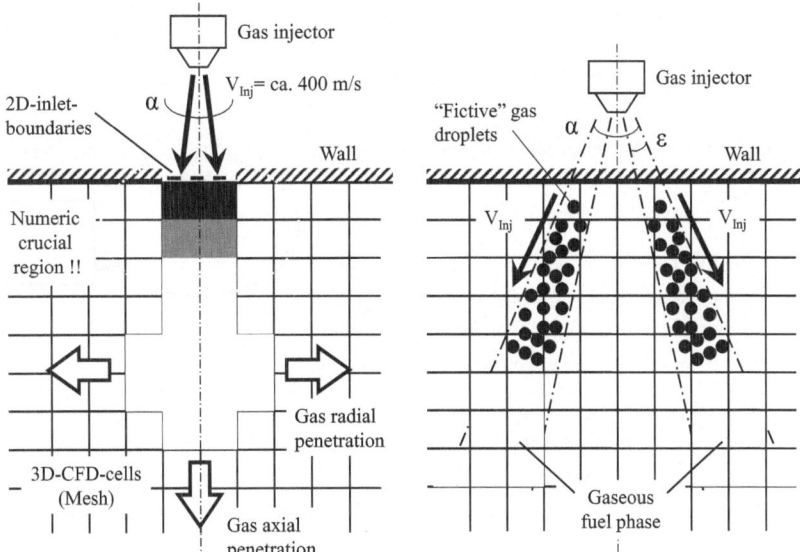

Figure 11.5: *Traditional gas injection modeling.*　　　　**Figure 11.6:** *Gas injection modeling in "QuickSim" (e.g. hollow jet cone).*

The space where the droplets are newly introduced embraces more cells than the few border cells involved in the traditional gas injection modeling (numeric crucial region), so that the high velocity gradients generated by the momentum transfer can be better distributed and solved in the near nozzle region. That means the 3D-CFD-simulation runs more stable. The size D of the "fictive" gas droplets is determined by a function that, in case of supersonic flow, takes the dimension of the Mach disk [98] into account. The initial calculated size D is approx. 1.5 times the orifice diameter and is kept constant. That means the modeling of break up phenomena due to aerodynamic forces required for liquid fuel is switched off in this case. Both initial density and temperature of the injected methane bubbles are equal to the conditions at the injector nozzle. When the "fictive" gas droplets (bubbles) reach an adequate distance from the nozzle (ca. 5-10 mm), so that the building of an accurate spatial jet in terms of the standard Eulerian formulation is ensured, a conversion of the methane bubbles into a continuum fluid takes place. During this "fictive" evaporation process, in comparison to liquid fluid, the influence of the heat of vaporization is also switched off. Thanks to this approach, the simulation of complex jet developments also using coarse meshes is possible.

11.5 3D-CFD-Domains limited to the Cylinder

The procedure here discussed represents the most simple and CPU-time efficient approach for the simulation of a cylinder (in this project: cylinder 3). The following investigations using two different extensions of the discretized domains have been performed:

- The fuel injectors in the mesh are outside the discretization area, i.e. the fuel is provided to the mesh through the intake 1D-boundary condition (see Figure 11.7). In this case the experimental pressure traces of the intake manifold (located between the injectors and the airbox) cannot be used as boundary conditions.

- The discretization of the intake manifold is extended nearly up to the airbox (see Figure 11.11), i.e. the location of the fuel injectors are part of the mesh and the injection can be adequately simulated (see Chapter 11.4). The location of the 1D-boundary condition is given by the position of the low-pressure sensors at the test bench.

11.5.1 3D-CFD-Simulation excluding the Fuel Injectors

In this approach the boundary conditions at the intake manifold are provided by a calibrated 1D-CFD-program. The location of the boundary condition in the exhaust manifold corresponds to the location of the low pressure sensor, so that its pressure trace can be directly used in the simulation (see Figure 11.7).

The fuel is passed to the mesh directly and homogeneously over the surface of the intake boundary condition by means of data from 1D-CFD-programs or estimations that help defining the scalar mass fractions of the input file (in this project values from 1D-CFD-simulations have been used). In this approach the influence of the fuel injection on the mixture formation and the momentum variation of the fresh charge is not reproducible (in case of an engine with liquid injection also the influence of the heat of vaporization on the charge temperature cannot be reliably reproduced). At the location of the intake boundary condition, where remarkable mixing processes take place, inconsistencies in the species mass balances often occur in case of strong backflows (see Chapter 11.3.2). For this reason, at least, the amount of intake fuel in the cylinder has to be controlled and eventually adjusted using correction factors for the species mass fraction of the boundary condition input file.

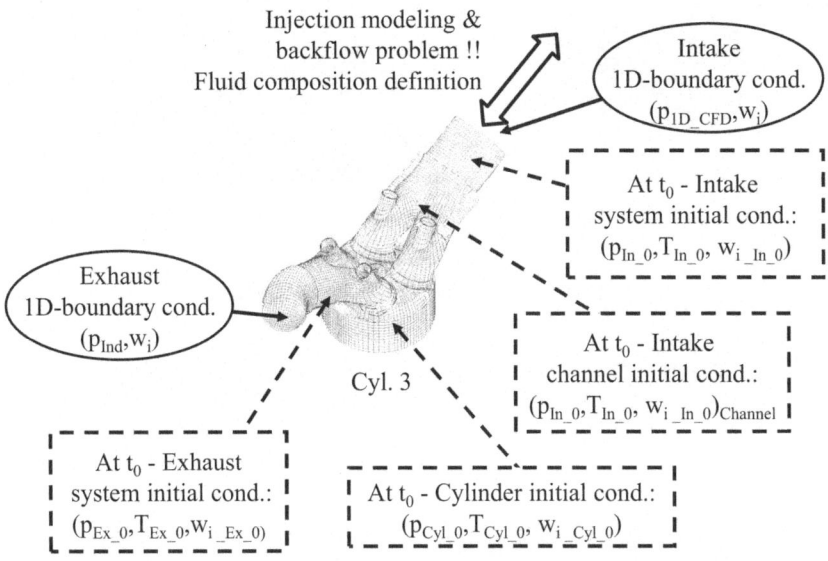

Injection modeling &
backflow problem !!
Fluid composition definition

Intake
1D-boundary cond.
(p_{1D_CFD}, w_i)

At t_0 - Intake
system initial cond.:
$(p_{In\_0}, T_{In\_0}, w_{i\_In\_0})$

Exhaust
1D-boundary cond.
(p_{Ind}, w_i)

At t_0 - Intake
channel initial cond.:
$(p_{In\_0}, T_{In\_0}, w_{i\_In\_0})_{Channel}$

Cyl. 3

At t_0 - Exhaust
system initial cond.:
$(p_{Ex\_0}, T_{Ex\_0}, w_{i\_Ex\_0})$

At t_0 - Cylinder initial cond.:
$(p_{Cyl\_0}, T_{Cyl\_0}, w_{i\_Cyl\_0})$

Figure 11.7: 3D-CFD-domain of the cylinder excluding the fuel injectors.

Figure 11.8 shows the fuel distribution (mass fraction) during the intake phase in the 3rd simulated cycle and after the adjustment of the amount of inflowing fuel to the target lambda-value at the ignition point. As mentioned before the influence of the injectors on the mixture formation is missing. Since the fuel inserted into the mesh using 1D-CFD-boundary conditions has a time-dependent fuel mass-fraction $w_F(\varphi)$, that tries to reproduce the profile of the injector mass flow, the fresh charge is not premixed to a fixed lambda-value. That means mixing processes among residual gas and fresh charge regions with different lambda-values take place up to the end of the combustion. Figure 11.9 shows the lambda distribution at the ignition point (IP=25 deg before FTDC – 5500 rpm - WOT). Here the mixture is well homogenized but, as further investigations will show (see Chapters 11.5.2-11.7), this is the result of an overestimation caused by the approach of fuel insertion through the intake 1D-boundary condition instead of the implementation of an injector-model.

Intake Cylinder 3 (Cyl_without_Inj.)

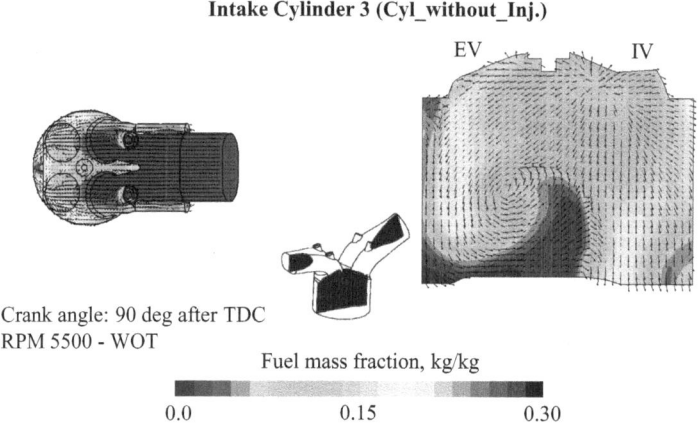

Crank angle: 90 deg after TDC
RPM 5500 - WOT

Fuel mass fraction, kg/kg

0.0   0.15   0.30

Figure 11.8: Fuel mass fraction and flow field during the intake phase
- 90 deg after TDC - 5500 rpm - WOT (cylinder mesh without injectors – 3rd cycle).

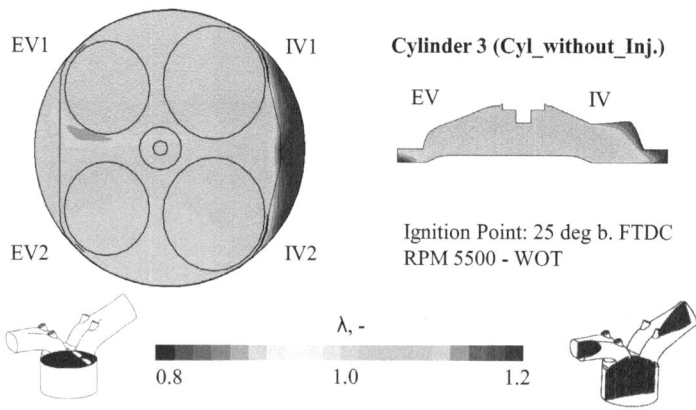

Ignition Point: 25 deg b. FTDC
RPM 5500 - WOT

λ, -

0.8   1.0   1.2

Figure 11.9: Lambda distribution at the ignition point
- IP=25 deg before FTDC - 5500 rpm - WOT (cylinder mesh without injectors – 3rd cycle).

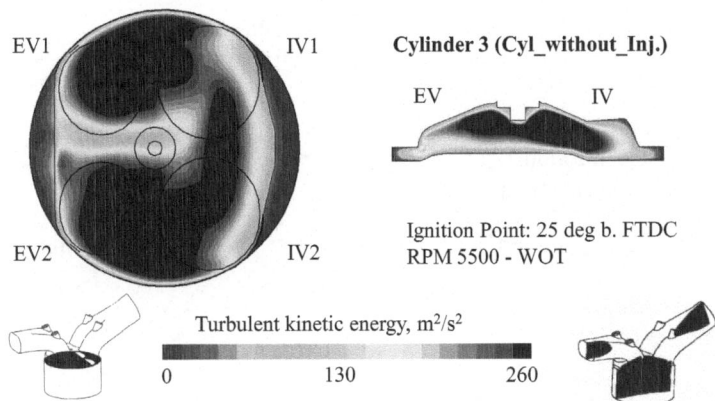

Figure 11.10: *(TKE) Turbulent kinetic energy distribution at the ignition point*
- IP=25 deg before FTDC - 5500 rpm - WOT (cylinder mesh without injectors – 3rd cycle).

The lambda distribution like the distribution of the specific turbulent kinetic energy (TKE) is approximately symmetric (see Figures 11.9 and 11.10). Even in case of a completely symmetric geometry of the mesh at the tumble-plane at any time, small numerical differences in the solution of the flow field have to be expected, so that at the end a perfectly symmetric solution cannot be reached. In contrast to the real engine where a higher degree of asymmetry is expected, here the effect of the airbox geometry on the in-cylinder flow motion is reproduced only by intake 1D-CFD-boundary conditions, i.e. eventual asymmetries over the surface of the boundary conditions cannot be taken into account.

The required CPU-time for the simulation of an operating condition (see Table 11.2) is very short also using only one processor (ca. 3 hours/cycle). Depending on the setting of the initial conditions (in particular the amount of fuel in the intake channels) the number of cycles required for reaching the convergence can slightly vary. Usually at WOT three cycles (sometimes also two cycles) are sufficient for a good convergence of the results.

11.5.2 3D-CFD-Simulation including the Fuel Injectors

This approach with pressure traces from the indicating system as manifold boundary conditions and with the implementation of the fuel injection modeling (see Figure 11.11) represents a remarkable improvement in comparison to the previous one (see Chapter 11.5.1).

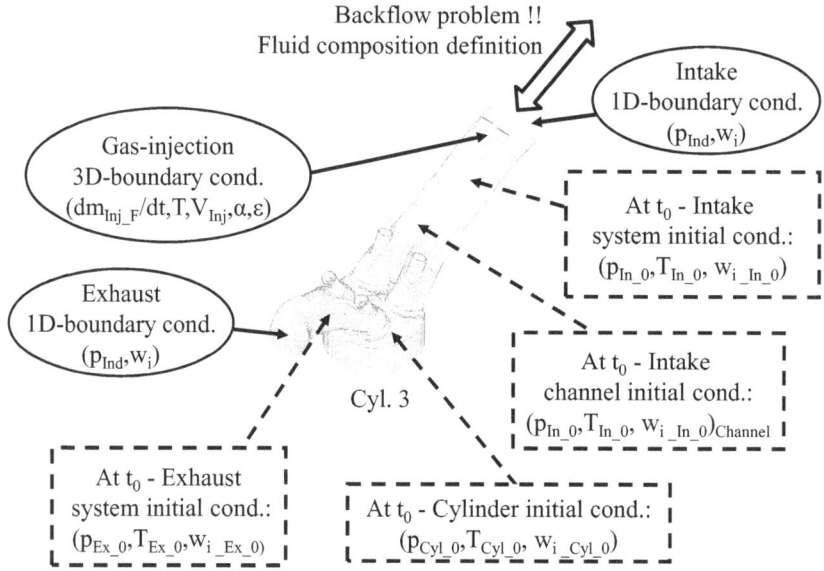

Figure 11.11: 3D-CFD-domain of the cylinder including the fuel injectors.

In this approach a more realistic mixture formation can be achieved. As shown in Figure 11.13 and well reported in the literature [88] the fuel does not mix in the intake channels because it builds a well defined stream along the wall. Only shear forces inside the combustion chamber, mainly generated by tumble and eventually swirl motion of the charge, support the mixing process.

Like in the previous approach (cylinder mesh without injectors), the problem of backflow in the intake channel still remains. When the gas leaves the 3D-CFD-domain (see Figure 11.12) the mass balance of the fuel is often compromised. Two methods help correcting this mass balance. The mass of injected fuel is adjusted to the target value of lambda at IP (more preferable way like in these simulations) or the fuel is reinserted in the mesh through the intake 1D-CFD-boundary condition whereas the local distribution at the boundary surface is compromised.

At the ignition point the lambda distribution (see Figure 11.14) shows, as expected, a remarkably lower degree of homogenization in comparison to the less realistic case of the mesh without injectors (see Figure 11.9). Since the distribution of the fuel at the injector nozzle is determined by a stochastic procedure there is no control towards a perfect symmetric injection. Therefore the mixture formation in this case is also affected by these asymmetries.

Cylinder 3 (Cyl_with_Inj.)

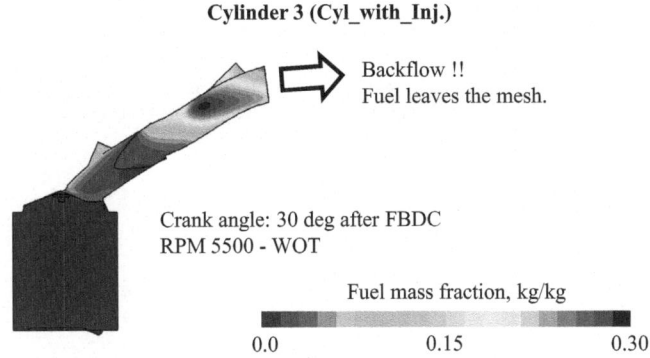

Figure 11.12: Fuel back flow out of the 3D-CFD-domain - 30 deg after FBDC - 5500 rpm - WOT (cylinder mesh with injectors – 3rd cycle).

Intake Cylinder 3 (Cyl_with_Inj.)

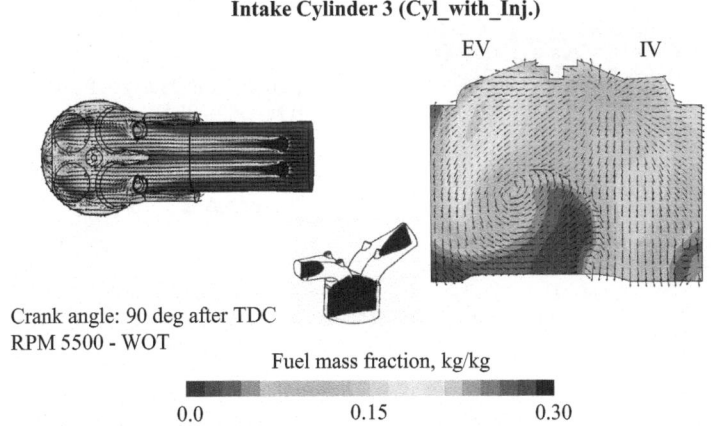

Figure 11.13: Fuel mass fraction and flow field during the intake phase - 90 deg after TDC - 5500 rpm - WOT (cylinder mesh with injectors – 3rd cycle).

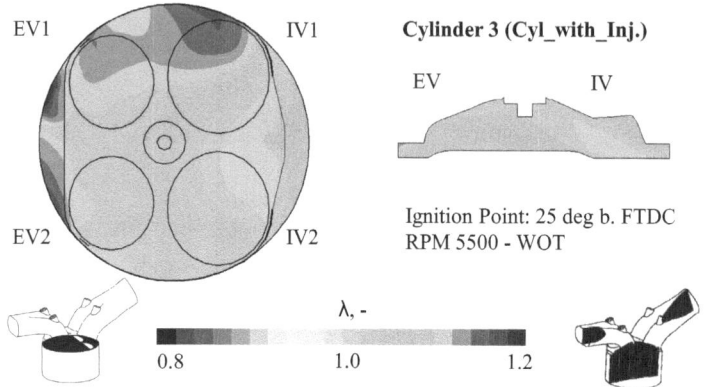

Figure 11.14: Lambda distribution at the ignition point
- IP=25 deg before FTDC - 5500 rpm - WOT (cylinder mesh with injectors – 3rd cycle).

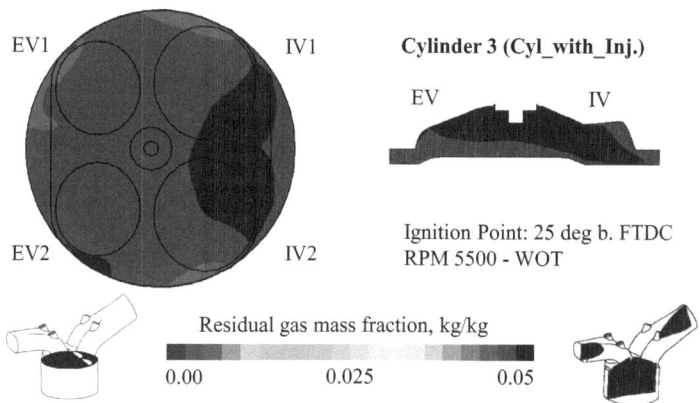

Figure 11.15: Residual gas mass fraction at the ignition point
- IP=25 deg before FTDC - 5500 rpm - WOT (cylinder mesh with injectors – 3rd cycle).

The residual gas distribution at the ignition point (see Figure 11.15), in contrast to lambda, shows a good homogenization within the charge. The average value of the residual gas ($w_{EGR}(\varphi_{IP}) \cong 5\%$) can be considered as a plausible value at the moment (more details in Chapter 11.7).

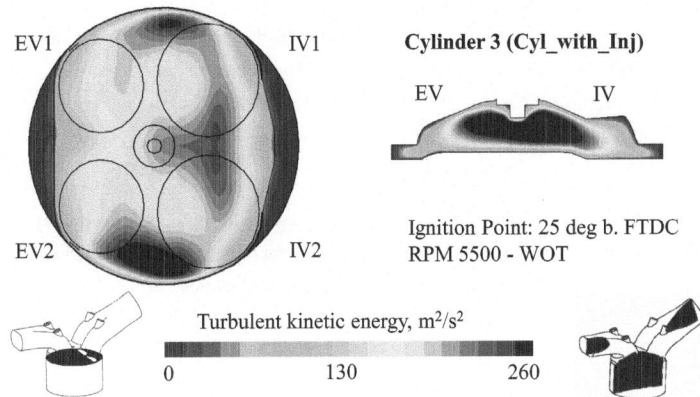

Figure 11.16: (TKE) Turbulent kinetic energy distribution at the ignition point
- IP=25 deg before FTDC - 5500 rpm - WOT (cylinder mesh with injectors – 3rd cycle).

The turbulence distribution in Figure 11.16 shows an impressive reduction in comparison to the previous approach (see Figure 11.10). This is mainly due to the distance between the intake 1D-CFD-boundary condition and the intake valves. Usually the lower the distance, the higher is the resulting turbulence level inside the combustion chamber. For this reason the best solution is always the simulation of the cylinder together with the airbox (see Chapters 11.6 and 11.7).

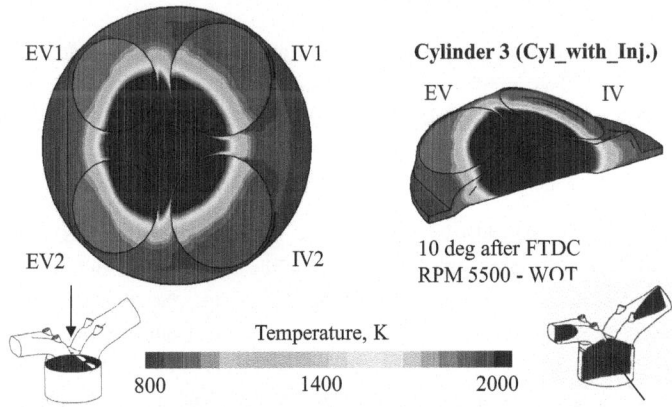

Figure 11.17: Temperature distribution during combustion
- 10 deg after FTDC - 5500 rpm - WOT (cylinder mesh with injectors – 3rd cycle).

The temperature distribution in Figure 11.17 shows the flame position at CA=10 deg after FTDC (burned mass fraction ca. 35%). The flame remains in the central position of the combustion chamber and still has a circular shape. The flame front characterized by a sharp temperature gradient has overall an approximately constant thickness, which usually covers one or two grid-cells (low numerical diffusion also using coarse meshes).

Like in the previous approach the required CPU-time for the simulation of an operating condition (see Table 11.2) is very short also using only one processor (ca. 3.5 hours/cycle). Also the number of required cycles for the convergence does not change, i.e. usually at WOT three cycles (sometimes also two cycles) are sufficient for a good convergence of the results.

11.6   Extension of the 3D-CFD-Domain: One Cylinder with the Airbox

The simulation of a cylinder together with the airbox allows recognizing the interactions between the flow within a detailed discretization of the airbox and the charge motion in the cylinder (see Figure 11.18). The charge flowing three-dimensionally through the airbox into the simulated cylinder is not disturbed by the setting of the boundary conditions and eventual backflows and complex mixing processes within the intake manifolds can be properly detected and reported.

In the proposed discretization of the airbox the domain ends near the throttle (excluded) and only small parts of the intake manifolds of the missing cylinders are included in the mesh. Conveniently the locations of the 1D-CFD-boundary conditions of the missing cylinders correspond to the locations of the pressure sensors, i.e. only the fuel injectors of the simulated cylinder are explicitly modeled (see Figure 11.18).

As mentioned before, a 3D-CFD-simulation of one cylinder with its airbox represents a remarkable advantage in comparison to the simulation of the cylinder alone. But with focus on a virtual engine development it has to be considered whether the simulated cylinder is representative for the whole engine or not. This point is very critical, otherwise more simulations (one for each cylinder) have to be performed and the correctness of the boundary conditions of the missing cylinders have to be verified. Only in this case the "sum" of all the simulation results can be considered consistent to the engine behavior. Therefore a virtual engine development based on this approach still remains very complicated.

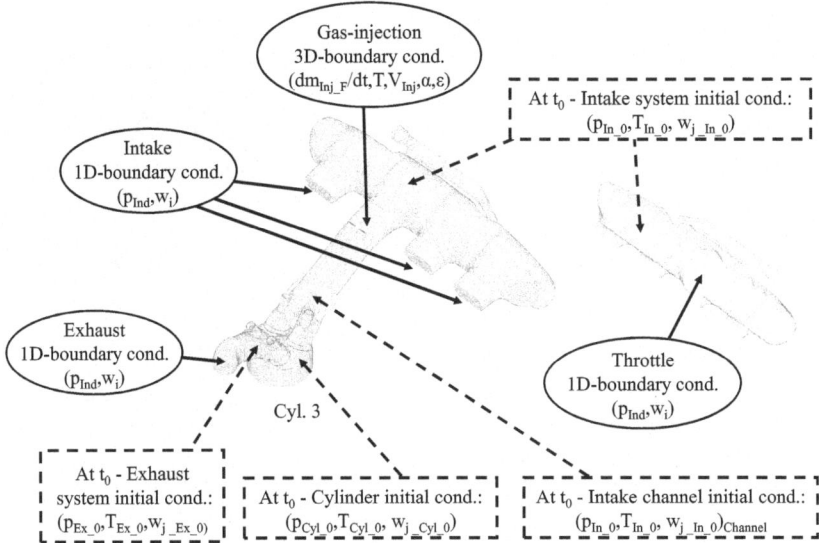

Figure 11.18: *3D-CFD-domain of one cylinder with the airbox.*

11.6.1 Between Predictability and Results Consistency

At the beginning of this approach using pressure trace signals (no velocity imposition – see Chapter 11.3.2) as boundary conditions for the missing cylinders, sometimes a curious phenomenon has been observed. In these cases despite the correct fulfilling of the pressure boundary conditions, a plausible mass flow through the throttle and a stable simulation-run the mass balances among the missing cylinders were completely out of convergence. This was caused by a recirculation between two intake boundary conditions (through one flowed three-times the normal mass and through the other flowed a negative mass), that irremediably compromised the results (see Figure 11.19). This problem occurs starting from engines with four cylinders and is more frequent by increasing the number of cylinders connected to the airbox.

The simplest solution to this problem is the setting of mass flows at the boundary conditions but as discussed in Chapter 11.3.2 this means a drastic reduction of the predictability of the tool, i.e. under these circumstances the tool is no longer able to find appropriately different results at many locations by varying e.g. the engine design.

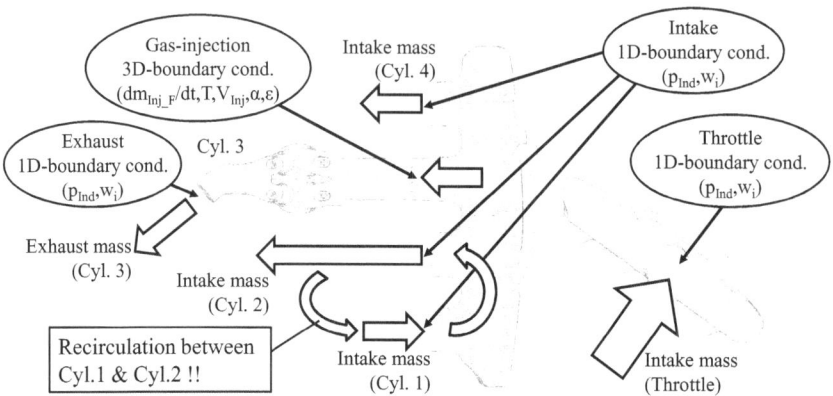

Figure 11.19: *A source of wrong results: the fluid recirculation between two or more cylinders.*

For this reason starting with the year 2007 an improved approach in the 3D-CFD-simulation of a cylinder with an airbox has been introduced in *QuickSim*.

11.6.1.1 *QuickSim*'s Improved Approach: The integrated 0D- and1D-CFD-Simulation of the missing Cylinders

In order to remove the problem introduced in Chapter 11.6.1 additional models implemented into *QuickSim* allow to simulate one-dimensionally the missing part of the intake manifolds of the other cylinders (see Figure 11.20). The missing parts are modeled as one-dimensional volumes in which mass and momentum conservation equations are applied. For this purpose the implemented models use also zero-dimensional boundary conditions at the intake valves extrapolated from the simulated 3D-CFD-cylinder. Good accuracy has always been achieved using only one volume for each missing manifold.

Figure 11.21 shows the fuel distribution in the intake manifold and within the airbox. Depending on the operating condition, a relevant amount of fuel can reach the airbox and also fuel swapping from cylinder to cylinder may occur. Since the fuel injectors of the missing cylinders are not included in the 3D-CFD-domain a realistic fuel mass fraction under backflow conditions has been inserted into the mesh using the 1D-CFD-boundary-conditions.

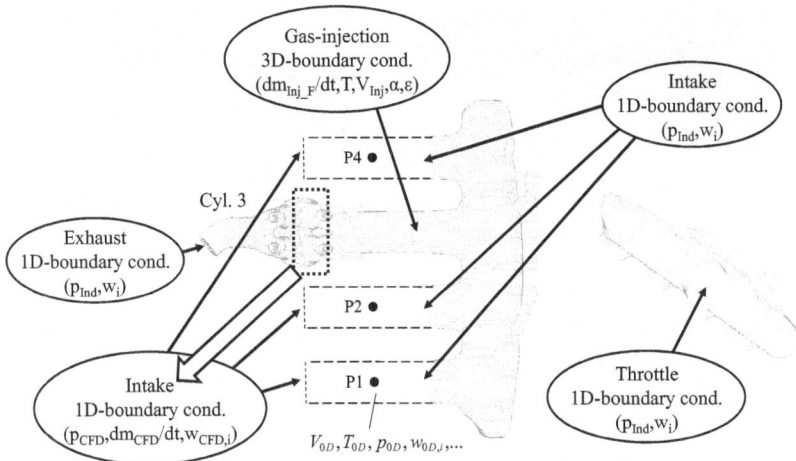

Figure 11.20: 3D-CFD-domain of one cylinder with the airbox and an integrated 0D- and 1D-CFD-simulation of the missing cylinders.

A remarkable asymmetry of the fuel mass fraction between the two intake ports is here evident. This is the effect of the flow field within the airbox that due to the position of the throttle and the chosen airbox design generates different mass flows between the intake valves. Therefore the resulting flow field is also able to slightly change the direction of the fuel jets with relevant effects on the mixture formation (see Figure 11.22). Here at the ignition point, in comparison to the simulations without airbox (see Figures 11.9 and 11.14) it becomes evident how sensitive are the results of a 3D-CFD-simulation with the extension of the discretized domain. In this approach, as it will be confirmed by the simulation of the full engine (see Figure 11.35), a more realistic mixture formation can be detected.

The turbulence distribution at the ignition point in Figure 11.23 shows an additional reduction of the turbulence level in comparison to the simulation with the cylinder alone. As mentioned in Chapter 11.5.2 this is mainly due to the distance between the intake 1D-CFD-boundary conditions and the intake valves (usually the lower the distance, the higher is the resulting turbulence level inside the combustion chamber). In this approach, as it will be confirmed by the simulation of the full engine (see Figure 11.41), also a more realistic turbulence distribution can be detected.

20 deg b. IVO - RPM 5500 - WOT

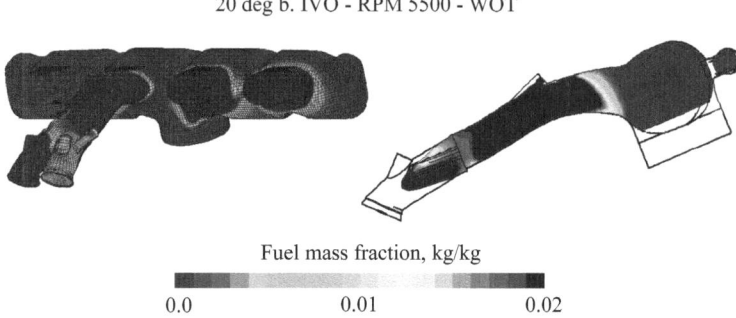

Fuel mass fraction, kg/kg

0.0   0.01   0.02

Figure 11.21: *Fuel mass fraction in the airbox*
- 20 deg before IVO - 5500 rpm - WOT (cylinder mesh with airbox – 7th cycle).

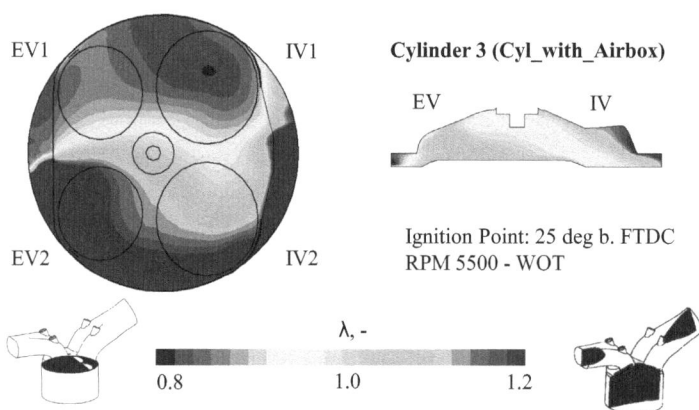

EV1   IV1   **Cylinder 3 (Cyl_with_Airbox)**

EV   IV

Ignition Point: 25 deg b. FTDC
EV2   IV2   RPM 5500 - WOT

λ, -

0.8   1.0   1.2

Figure 11.22: *Lambda distribution at the ignition point*
- IP=25 deg before FTDC - 5500 rpm - WOT (cylinder mesh with airbox – 7th cycle).

The required CPU-time for the simulation of an operating condition (see Table 11.2) is longer than in case of the simulation of the cylinder alone (ca. 18 hours/cycle). This is principally due to the greater number of cells in the mesh. Depending on the setting of the initial conditions (in particular the initial pressure and temperature of the airbox) the number of cycles required for reaching the convergence can vary remarkably. Usually at WOT and in case of a small airbox (like in this engine configuration), seven cycles (sometimes also five cycles) are sufficient for a good convergence of the results. For a bigger airbox the number of cycles required for convergence can easily increase up to 15-20 cycles.

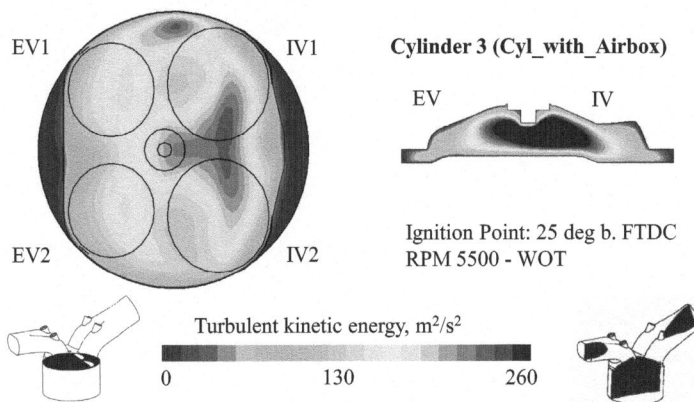

Figure 11.23: *(TKE) Turbulent kinetic energy distribution at the ignition point - IP=25 deg before FTDC - 5500 rpm - WOT (cylinder mesh with airbox – 7th cycle).*

11.7 3D-CFD-Simulation of the Full Engine

In this approach all the cylinders, the fuel injectors, the airbox starting from the throttle and finally the exhaust system up to the turbine entrance are included in the 3D-CFD-domain (see Figure 11.24). Only two 1D-CFD-boundary-conditions at the manifolds are required (intake throttle and turbine entrance). The extension of the 3D-CFD-domain in comparison to the previous cases allows investigating the behavior of each cylinder and their interactions with the airbox and the discretized exhaust system with high reliability. Mixture formation and all other relevant processes take place in the domain so that problems related to e.g. backflows and setting of boundary conditions (see Chapters 11.5 and 11.6) can be eliminated.

11.7.1 Results and 3D-CFD-Flow Field Investigations on the Full Engine

On the next pages some results of the full-engine simulation at 5500 rpm WOT are shown. The diagrams are provided automatically by the evaluation tool of *QuickSim*. Due to secrecy often the variable profiles are normalized to the maximum value of cylinder 1.

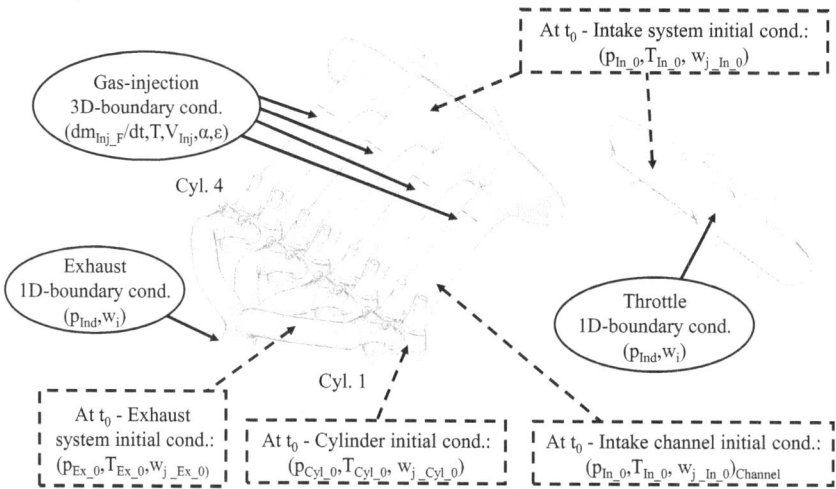

Figure 11.24: *3D-CFD-domain of the full engine.*

Analyzing the mass flow through the intake and exhaust valves (see Figures 11.25 and 11.26) it becomes evident how the airbox design influences the intake phase of each cylinder. The external cylinder 1 and 4 (blue lines) have a relatively sharp peak while the internal cylinders 2 and 3 (red lines) show a more flat maximum. This behavior is mainly due to the small volume of the airbox and the central position of the throttle manifold, that having a relatively constant mass flow reduces the pressure waves in the facing cylinders, i.e. the central ones. In contrast, the mass flow profiles through the exhaust valves show smaller differences among the cylinders. The trapped air mass in the central cylinders takes more profit from the conditions in the intake manifolds and is about 3% higher than in cylinder 1 and 5% higher than in cylinder 4 after IVC (see Figure 11.27).

The differences in the pressure profile during the combustion are evident (see Figure 11.28). This is an evidence that the combustion durations have reasonable variations not only among the cylinders but also from cycle to cycle (see Chapter 11.8). Other results in this chapter aim to explain this behavior.

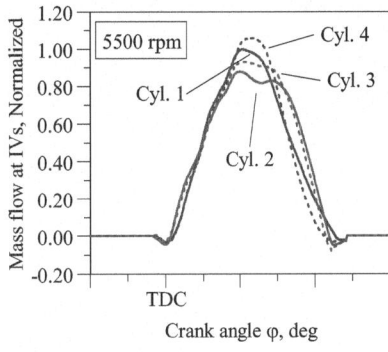

Figure 11.25: *Mass flow through IVs (mg/s) - 5500 rpm WOT – Full engine – 5ᵗʰ cycle.*

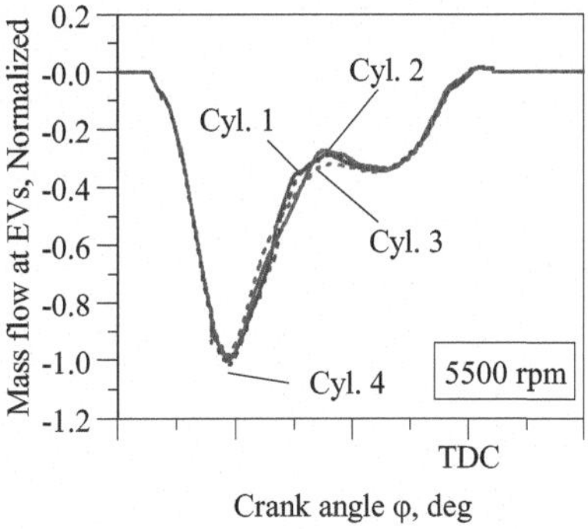

Figure 11.26: *Mass flow through EVs (mg/s) - 5500 rpm WOT – Full engine - 5ᵗʰ cycle.*

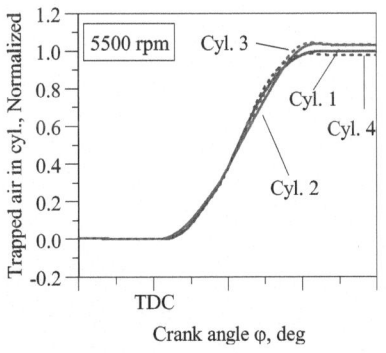
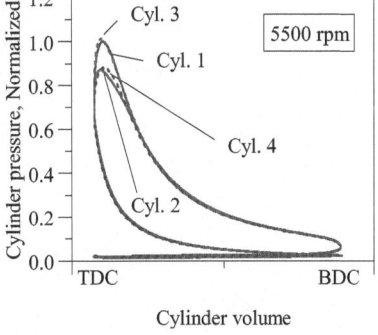

Figure 11.27: *Trapped air in cylinder (mg) - 5500 rpm WOT – Full engine - 5ᵗʰ cycle.*

Figure 11.28: *Cylinder pressure (bar) - 5500 rpm WOT – Full engine - 5ᵗʰ cycle.*

Also the tumble and swirl ratio profiles, respectively, show differences between the external and the central cylinders (see Figure 11.29 and 11.30). The central cylinder due to "smoother" intake phases have a remarkably lower tumble level up to the end of the compression stroke. The swirl profile, as expected, shows the direct influence of the airbox design on the flow motion within the cylinders. The charge flowing from the throttle manifold to the external cylinders gets a higher momentum that generates asymmetries with different mass flows through the intake valves and finally a higher swirl (the rotation direction of the swirl is in accordance with this consideration). The resulting swirl is in all cases very low but it can have great influence on the mixture formation of CNG engines.

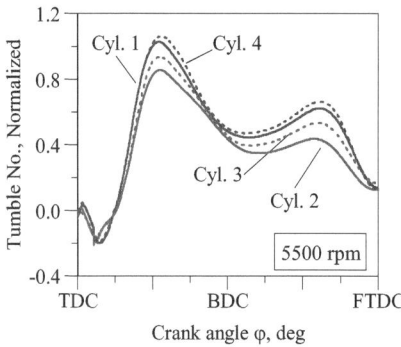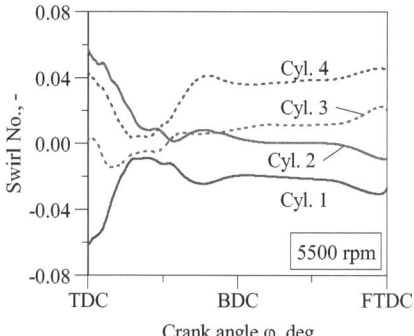

Figure 11.29: *Tumble ratio (-)*
- 5500 rpm WOT – Full engine - 5ᵗʰ cycle.

Figure 11.30: *Swirl ratio (-)*
- 5500 rpm WOT – Full engine - 5ᵗʰ cycle.

11.7.1.1 Mixture Formation

The goal of a homogeneous fuel-air mixture in CNG-engines is a very complex task. The fuel injectors are only able of a very low radial penetration (low spatial distribution – see Figure 11.33) and even an injection velocity of approx. 400 m/s generates also a poor axial penetration [88]. Actually after an axial penetration of ca. 25 mm the fuel does not have a relative velocity to the background fluid anymore. From this distance the fuel can be assumed to be transported by the motion of the charge. In Figures 11.31 and 11.32 the profiles of lambda as averages in the cylinder or locally at the spark plug are reported. These figures show that each cylinder reached the target fuel quantity ($\lambda \cong 1$) and also the differences among the spark plugs at the end of the compression stroke are small. These results would let suppose that the mixture in each cylinder is optimal, but, as further results will show, this is a wrong assumption (see Figure 11.35).

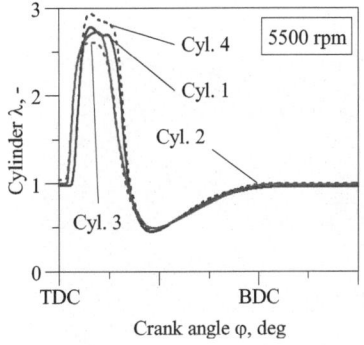

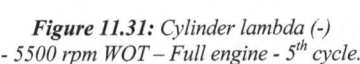

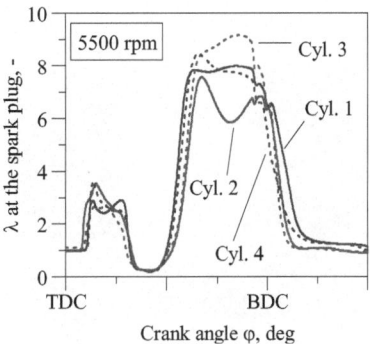

Figure 11.31: *Cylinder lambda (-)*
- 5500 rpm WOT – Full engine - 5ᵗʰ cycle.

Figure 11.32: *Lambda at the spark plug (-)*
- 5500 rpm WOT – Full engine - 5ᵗʰ cycle.

The mixture formation of this race engine at WOT can be described as follows. The high amount of fuel injected, also using two injectors for each cylinder, requires much longer time then the duration of the intake valve opening. Therefore a reasonable amount of fuel (usually up to 70% and even more for high distances between the injectors and the intake valves) remains always captured in the intake manifolds during closed valves (see Figure 11.33). Here, due to the time at disposal, pressure waves, flow asymmetries, etc., the fuel builds a quite homogenous rich mixture ("pre-load-mixture") with the charge. This mixture is then aspirated in the cylinder during the next intake stroke in addition to the actual injected fuel, which, in contrast, is characterized by a very worse homogenization with the charge in the manifolds (drawback of gas injection). In this case it has been observed that the flow asymmetries in the intake manifolds help the homogenization of the "pre-load-mixture" (cylinders 1 and 4 – see Figure 11.33) but, as already discussed in this chapter, they negatively influence the directions of the fuel jets of the actual injections, causing remarkable differences between the fuel mass flows through the valves. This means, that depending on the amount of the fuel in the "pre-load-mixture" and on the distance of the injectors to the valves (the longer the better) the effect of these flow asymmetries can be positive (like at these operating condition – see Figure 11.35).

At the end the mixture formation takes place in the combustion chamber (see Figure 11.34). Tumble and piston motion in combination with an appropriate combustion chamber design (no death-zones) are then responsible for the final fuel distribution. The swirl level in this engine has no influence on the mixture formation directly in the cylinder, i.e. the charge rotation within the cylinder is too small for an eventual correction of fuel inhomogeneities.

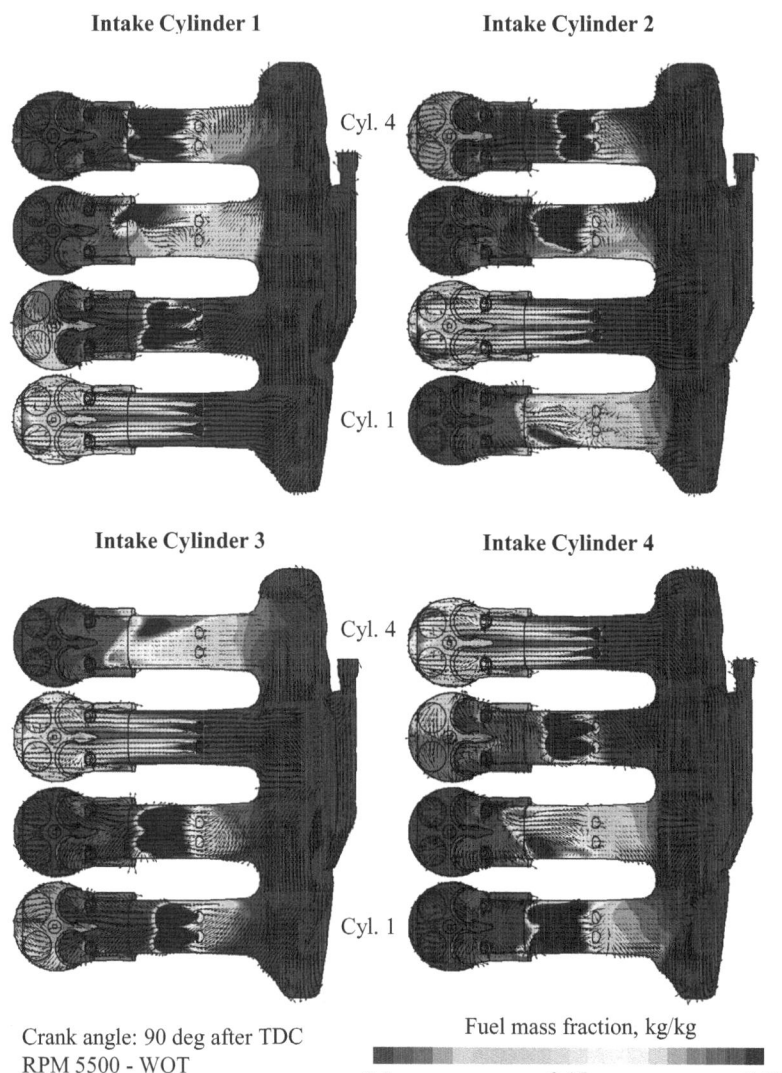

Figure 11.33: *Fuel mass fraction and flow field during the intake phase of each cylinder - 90 deg after TDC (cylinder relative) - 5500 rpm - WOT (full engine mesh – 5th cycle).*

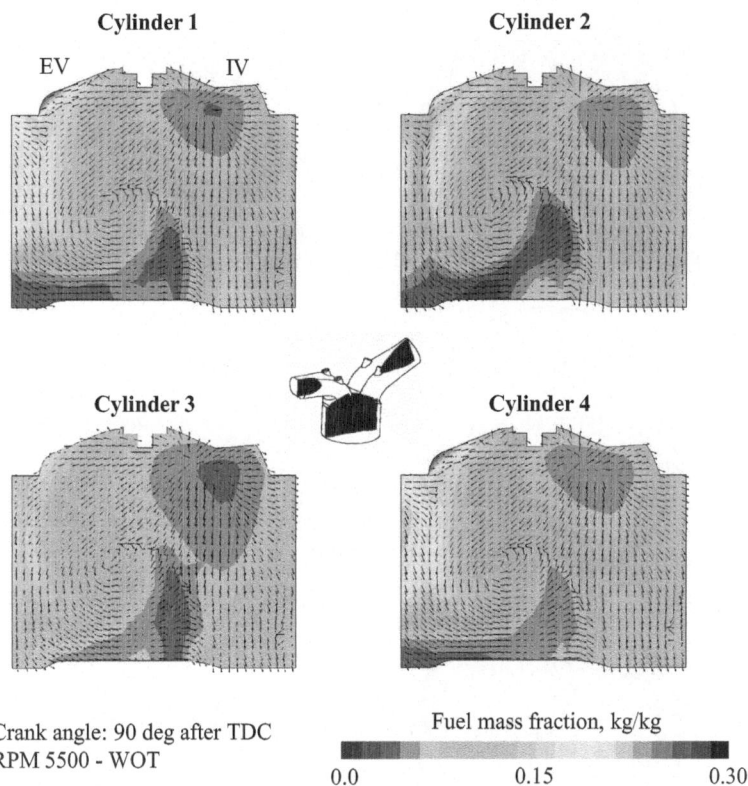

Cylinder 1

EV   IV

Cylinder 2

Cylinder 3

Cylinder 4

Crank angle: 90 deg after TDC
RPM 5500 - WOT

Fuel mass fraction, kg/kg

0.0   0.15   0.30

Figure 11.34: *Fuel mass fraction and flow field during the intake phase of each cylinder - 90 deg after TDC (cylinder relative) - 5500 rpm - WOT (full engine mesh – 5ᵗʰ cycle).*

Figure 11.35 shows the resulting mixture formation at the ignition point. The external cylinders are characterized by a good homogenization while the internal cylinders have a much worse distribution. In particular it is evident that the resulting distribution is symmetric between both the internal and external cylinders (relevant influence of the airbox design on the mixture formation).

The results presented here for the cylinder 3 are in accordance with the simulation results using the mesh of one cylinder with the airbox (see Chapter 11.6), confirming, in most of the cases, the importance of this approach.

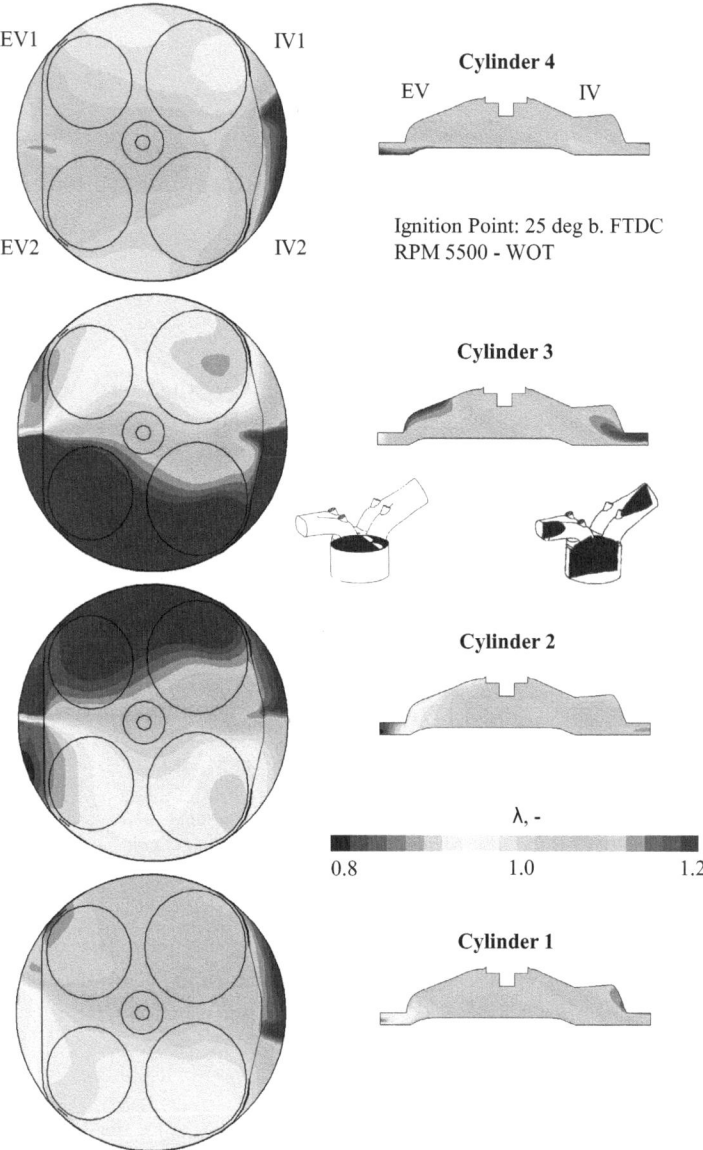

Figure 11.35: *Lambda distribution at the ignition point of each cylinder - IP=25 deg before FTDC (cylinder relative) - 5500 rpm - WOT (full engine mesh – 5th cycle).*

11.7.1.2 Residual Gas Distribution

Figures 11.36 and 11.37 show the profile of the residual gas mass fraction as averages over the cylinder and locally at the spark plug. The amount of residual gas after IVC is practically the same among the cylinders. As wished for race engines the concentration of the residual gas at WOT is quite low ($w_{EGR}(\varphi_{IP}) \cong 3\%$). This value is in accordance to a turbo-charged race engine with a small valve "overlap". Obviously the valve overlap towards an additional reduction of the residual gas concentration can be increased, but this is not a recommendable approach for CNG-engines.

Also the more sensitive concentrations of residual gas at the spark plug show very small differences among the cylinders and the final values at the ignition point are similar to the cylinder-averaged value. The profile $w_{EGR,SP}$ is common in all engines. First a rapid "scavenging" of the spark plug with the fresh charge mass-flow through the valves takes place during the intake phase and then successively due to the tumble motion of the charge and the related mixing processes the final value of concentration is reached.

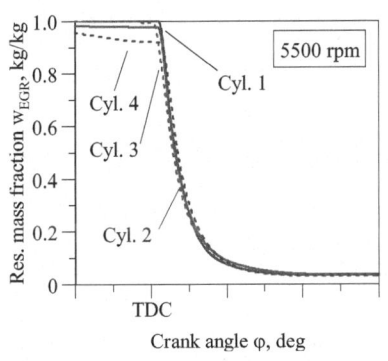

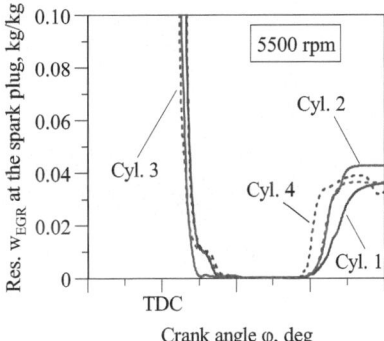

Figure 11.36: *Cylinder EGR mass fraction w_{EGR} (kg/kg) - 5500 rpm WOT – Full engine - 5th cycle.*

Figure 11.37: *EGR mass fraction at the spark plug $w_{EGR,SP}$ (kg/kg) - 5500 rpm WOT – Full engine - 5th cycle.*

The homogenization degree of the residual gas among the cylinders is very high (see Figure 11.38), that indicates a very effective scavenging process (this point is often critical in race engines). In comparison to the simulations without airbox (see Chapter 11.5) the concentration of residual gas is here lower (more accurate simulation of the intake phase – see Chapter 11.9).

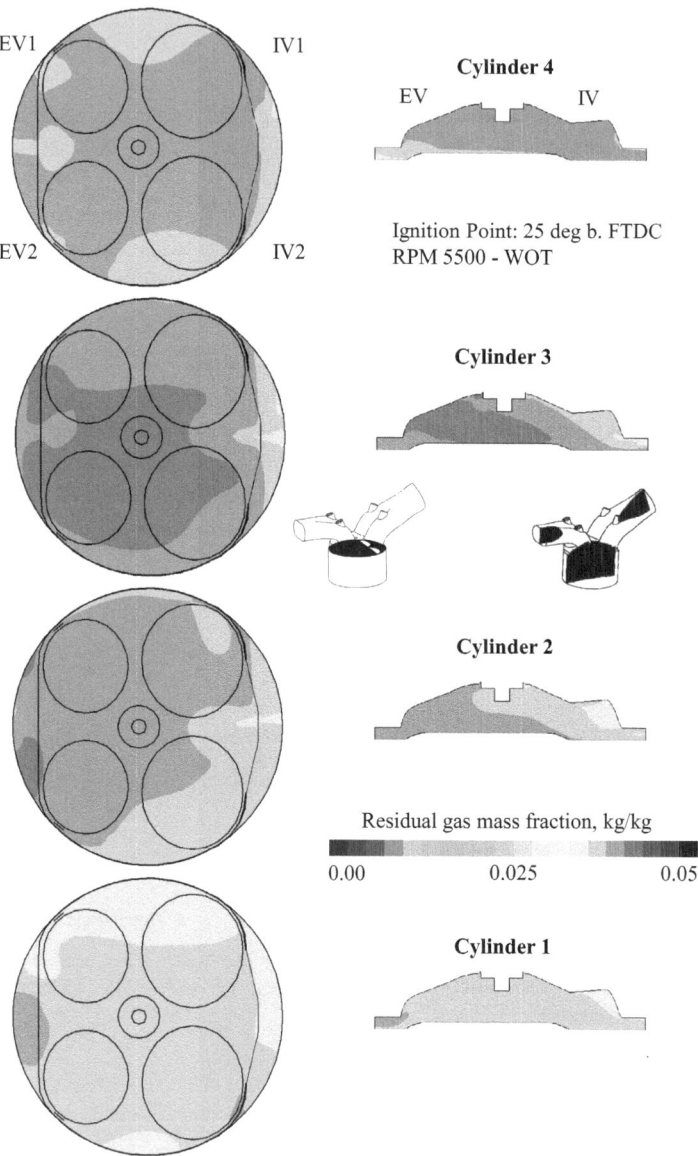

Figure 11.38: Residual gas mass fraction at the ignition point of each cylinder - IP=25 deg before FTDC (cylinder relative) - 5500 rpm - WOT (full engine mesh – 5th cycle).

11.7.1.3 **Turbulence**

The profiles of the turbulence as averages in the cylinder and local at the spark plug (see Figure 11.39 and 11.40) follow the trends shown by both the intake mass flow and the tumble ratio (see Chapter 11.7.1.1). Here the focus is mainly on the turbulence profile during the combustion.

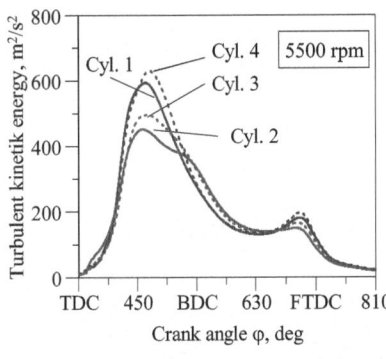
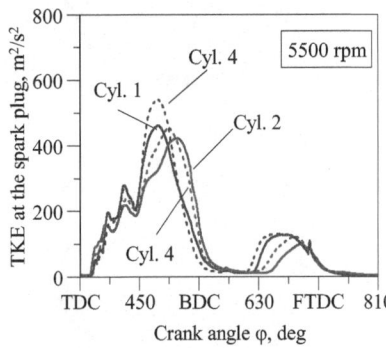

Figure 11.39: (TKE) – Cylinder turbulence - 5500 rpm WOT – Full engine - 5th cycle.   *Figure 11.40: Turbulence at the spark plug - 5500 rpm WOT – Full engine - 5th cycle.*

The external cylinders (blue lines), as expected, have a remarkably higher turbulence level during the intake stroke generated directly by the velocity gradients of the intake mass flow as soon as it has reached the combustion chamber (deceleration of the charge). Depending on the shape of the combustion chamber, due to the momentum conservation law, the charge starts rotating on the tumble axis up to its break down at the end of the compression stroke. This effect that generates a considerable dissipation of kinetic energy into turbulence can be seen in particular for the external cylinders in Figure 11.39 ($\varphi \cong 700$ deg – i.e. approximately at the ignition point). The internal cylinders show a "smoother" and earlier tumble break down during the compression stroke with very moderate turbulence increasing near FTDC. During the combustion, in contrast to the turbulence during the intake stroke, the differences among the cylinders and also locally at the spark plugs are smaller (see Figure 11.41). As visible in the pictures of the turbulence distribution within the combustion chamber, the external cylinders still have a moderate higher turbulence level (highly symmetric result between external cylinders).

In comparison to the simulations without airbox, that are drastically influenced by the location of the intake boundary condition (see Chapter 11.5), it is here evident how the presence of the airbox in the 3D-CFD-domain reduces the level of turbulence to more reliable results in both simulations: one cylinder with airbox (see Figure 11.23) or full engine (see Figure 11.41).

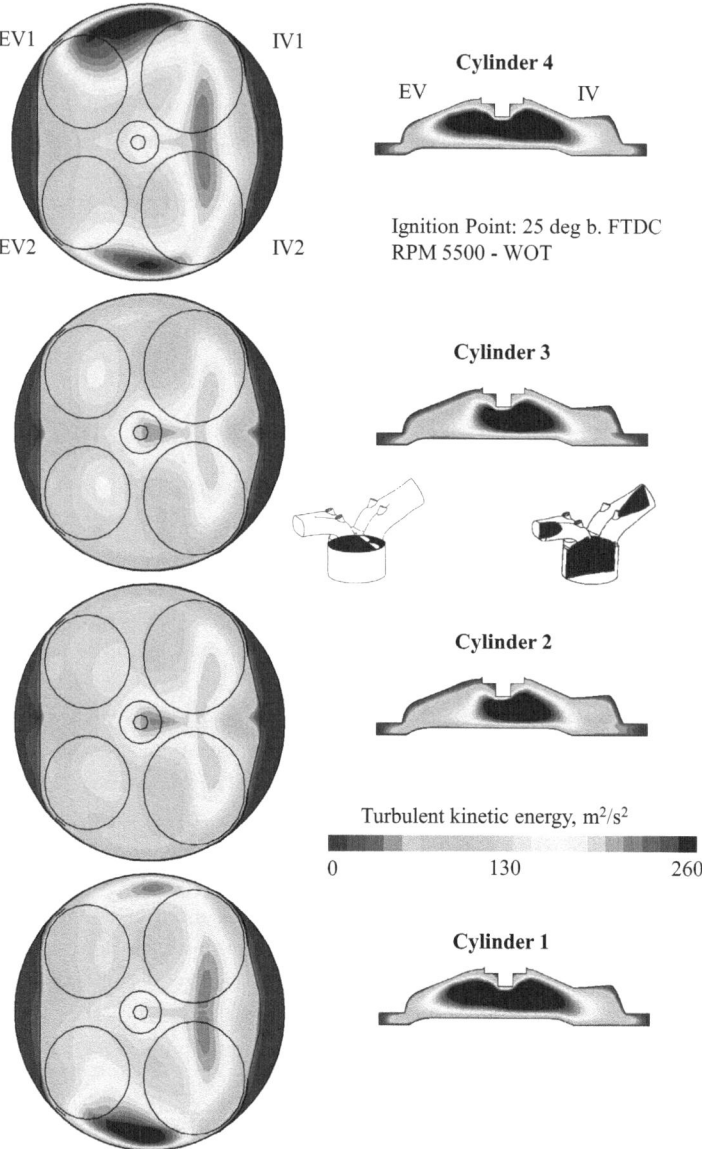

Figure 11.41: (TKE) Turbulent kinetic energy distribution at the ignition point of each cylinder - IP=25 deg before FTDC (cylinder relative) - 5500 rpm - WOT (full engine mesh – 5th cycle).

11.7.1.4 Combustion

As an outlook for combustion investigations, the analysis of the results reported in the previous paragraphs shows that the external cylinders have a better mixture homogenization and a higher turbulence level at the ignition point. In contrast the internal cylinders have a higher volumetric efficiency. Since experimental measurements have detected a quite "irregular" operation of cylinder 3 characterized by much higher cycle-to-cycle variations in comparison to the other cylinders, here a short analysis of the combustion process aims to conclusively find an explanation for this behavior.

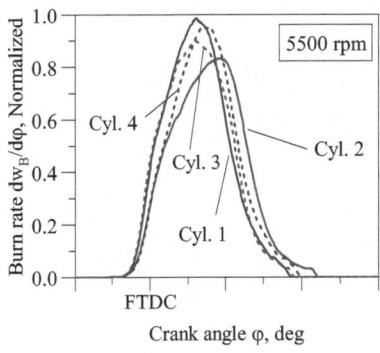

Figure 11.42: *Cylinder burn rate (%/deg) - 5500 rpm WOT – Full engine - 5th cycle.*

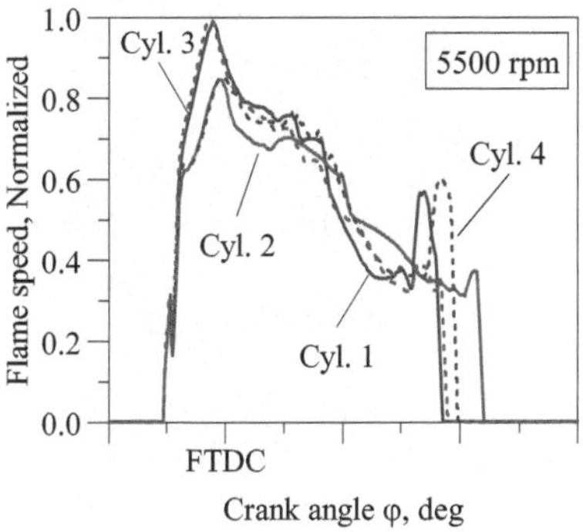

Figure 11.43: *Abs. flame speed (m/s) - 5500 rpm WOT – Full engine - 5th cycle.*

Figures 11.42 and 11.43 show the burn rate profile and for a deeper insight also the absolute flame propagation u_f (both normalized) during the whole combustion process. Cylinder 3 is surprisingly characterized by a very high flame speed up to its peak of burn rate. But despite the initial high flame speed the peak of burn rate is, as expected due to a worse homogeneous mixture (like in cylinder 2), lower than in the external cylinders. Here an answer can be found in the analysis of the turbulence profile of cylinder 3, that shows, in comparison to cylinder 2, higher gradients during the combustion period (see Figures 11.39 and 11.40).

From my experience in 3D-CFD-simulations the combination of a modest homogenization and high turbulence gradients are principally responsible for a very sensitive combustion process that amplifies (in terms of imep) the already present and remarkable cycle-to-cycle amplitude variations concerning the charge (see Chapter 11.8). These effects are particularly intense in cylinder 3 followed by cylinder 2 and much more moderate in the external cylinders.

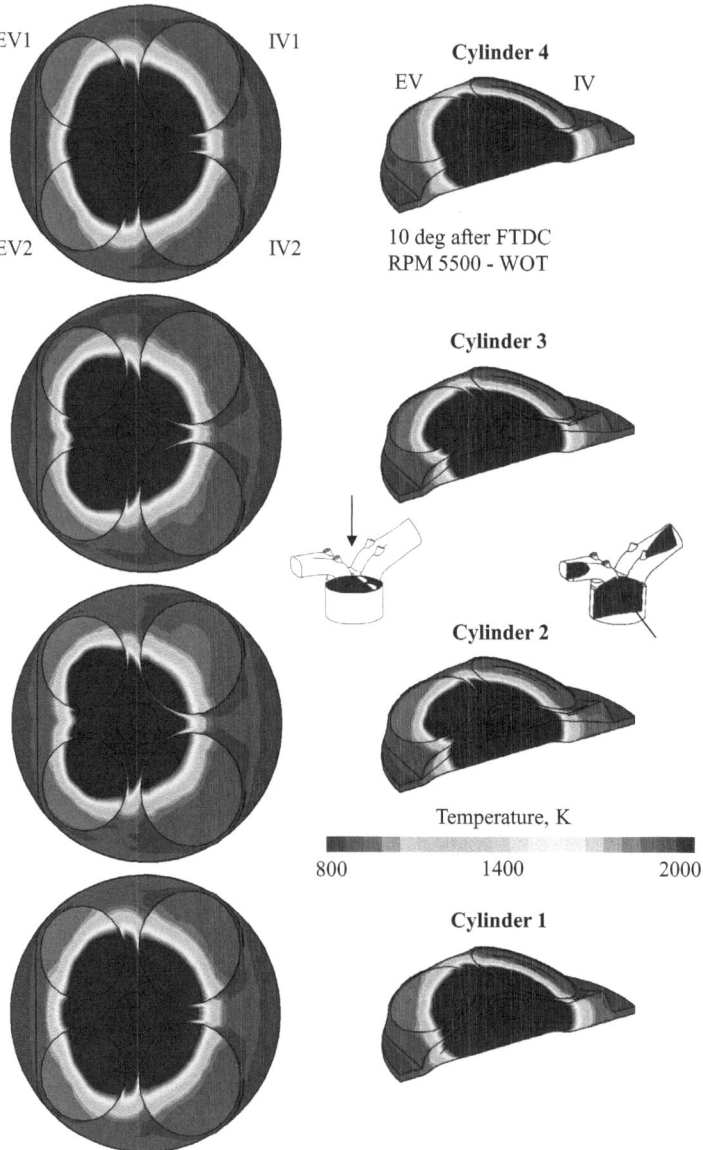

Figure 11.44: *Temperature distribution during combustion of each cylinder - 10 deg after FTDC (cylinder relative) - 5500 rpm - WOT (full engine mesh – 5th cycle).*

The temperature distributions in Figure 11.44 show the flame positions at CA=10 deg after FTDC (burned mass fraction ca. 35%). The flames in the external cylinders remain in the central position of the combustion chamber and still have a circular shape. In contrast the centers of the burned zones in the internal cylinders have moved near the exhaust valves and have a less circular shape (indication of a more sensitive combustion). In comparison the temperature distribution in Figure 11.17 (simulation of the cylinder alone with injectors) is not able to recognize this trend.

11.7.1.5 Convergence of the Results

The required CPU-time for the simulation of an operating condition (see Table 11.2) is only 30% longer than in case of the simulation of one cylinder with airbox (ca. 24 hours/cycle). Considering the dimension of the 3D-CFD-domain (450,000 vs. 190,000 cells) this is a very moderate CPU-time increasing. The reason of this result has to be found in a much more stable calculation (principally due to 1D-CFD-boundary conditions located in less sensitive regions, where velocity gradients are much lower) that permits to run the simulation with less iterations (time step $\Delta t \cong 0.5 \deg$) and less convergence corrector-steps within each time step. Depending on the setting of the initial conditions (in particular the initial pressure and temperature of the airbox) the number of cycles required for reaching the convergence can vary remarkably. Usually at WOT and in case of a small airbox (like in this engine configuration), five cycles are sufficient for a good convergence of the results. That means a full-engine simulation requires less simulation cycles to converge then the case with one cylinder with airbox (here a few cycles required for the convergence of the integrated 0D- and 1D-model for the simulation of the missing cylinder are skipped – see Chapter 11.6.1.1), so that at the end the total CPU-time is not longer then in the previous case.

11.7.2 Result Comparison among different Operating Conditions

On the next pages some results of the full-engine simulations at three different engine speeds (5500, 6000 and 6500 rpm WOT) are shown. The results refer to cylinder 3 and due to secrecy often the variables are normalized to the maximum value of the case at 5500 rpm.

The maximal intake mass flow of the engine is limited by both an air-restrictor and a pressure regulation in the airbox (see Chapter 11.2), i.e. as shown in Figure 11.46 the maximal amount of trapped air in the cylinder is reached between 5500 and 6000 rpm. At 6500 rpm, due to these limitations, the volumetric efficiency remarkably decreases. The profiles of the intake mass flow through the valves (see Figure 11.45) show moderate oscillations in all the investigated cases, which, as expected, slightly rise with increasing the engine speed. The mass flow peak is approximately the same and a small backflow at IVC is present only at 5500 rpm. From the mass flow diagram it can be seen that the intake valve timing has been optimized for a target engine speed at about 6000 rpm. The tumble ratio does not change remarkably, i.e. the rotation speed of the charge varies quite linearly with the engine speed (see Figure 11.47). At 6500 rpm the tumble ratio takes principally profit from an earlier cylinder filling (the mass flow profile is oriented towards the valve opening phase) that lets rise its value during the whole compression stroke. The turbulence in the cylinder, as usual, rises with increasing the engine speed (see Figure 11.48).

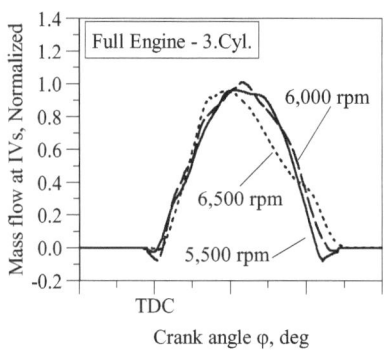

Figure 11.45: *Mass flow through IVs (mg/s) - Full engine (3. Cyl.) – 5th cycle – 5500/6000/6500 rpm WOT.*

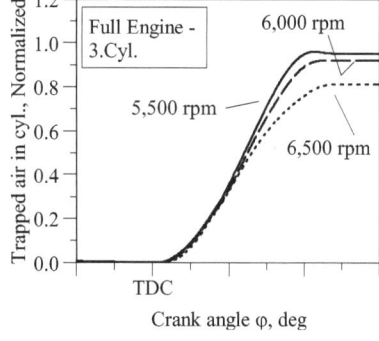

Figure 11.46: *Trapped air in cylinder (mg) - Full engine (3. Cyl.) – 5th cycle – 5500/6000/6500 rpm WOT.*

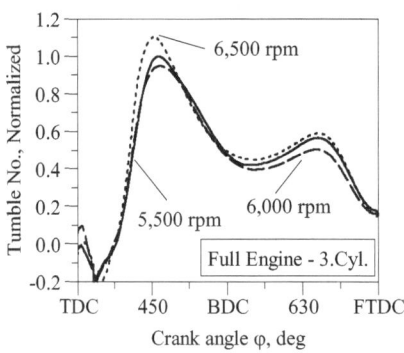

Figure 11.47: *Tumble ratio (-) - Full engine (3. Cyl.) – 5th cycle – 5500/6000/6500 rpm WOT.*

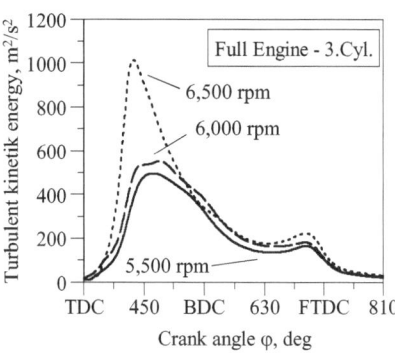

Figure 11.48: *(TKE) turbulence in cyl. (m^2/s^2) - Full engine (3. Cyl.) – 5th cycle – 5500/6000/6500 rpm WOT.*

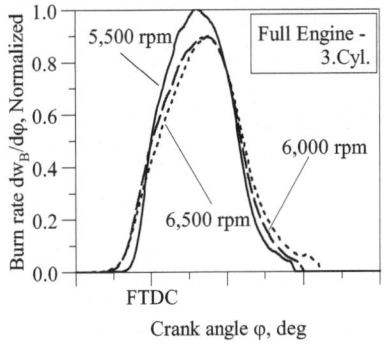

Figure 11.49: *Burn rate (%/deg) - Full engine (3. Cyl.) – 5ᵗʰ cycle – 5500/6000/6500 rpm WOT.*

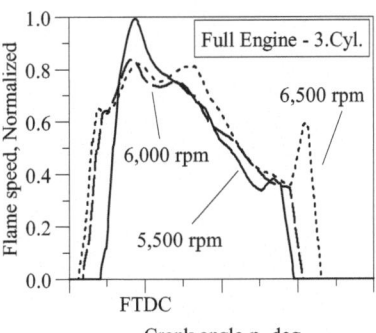

Figure 11.50: *Abs. flame speed (m/s) - Full engine (3. Cyl.) – 5ᵗʰ cycle – 5500/6000/6500 rpm WOT.*

The burn rate has a "steeply" profile at 5500 rpm and becomes smoother with increasing the engine speed (see Figure 11.49). Up to FTDC the flame speed is higher at 5500 rpm while in the second part of the combustion it is higher at 6500 rpm (see Figure 11.50). At the end, as usual, the combustion duration (expressed in degrees) rises with increasing the engine speed.

11.8 The Simulation of successive Operating Cycles

The results of a 3D-CFD-simulation can only be reliable if the flow motion within the 3D-CFD-domain reproduces the results within a certain convergence criteria over many simulated cycles. As introduced in Chapter 7.2.1, the simulation of only one cycle is not representative for the engine operation. Considering different extensions of the 3D-CFD-domain (see Chapters 11.5-11.7) and taking the tumble ratio (normalized) as a reference variable for these investigations, it can be seen that in no case the results of the first cycle have reached an acceptable convergence level (see Figures 11.51-11.54). All the simulations have been started using the same initial conditions and it was not possible to identify a clear trend for predicting the variations between the results of the first cycle and the successive one. Here, e.g. the simulation with the smallest extension of the domain (cylinder alone without injectors) shows a bigger difference from cycle 1 to cycle 2 then the simulation with the most extended domain (full-engine). The simulations without airbox reach the convergence at the third cycle while the simulation of the full engine has to run up to approximately the fifth cycle.

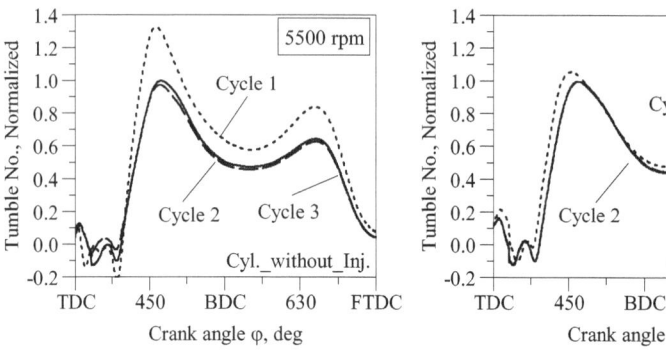

Figure 11.51: *Tumble ratio (-)*
- 5500 rpm WOT – Cylinder without injection -
Different simulation cycles.

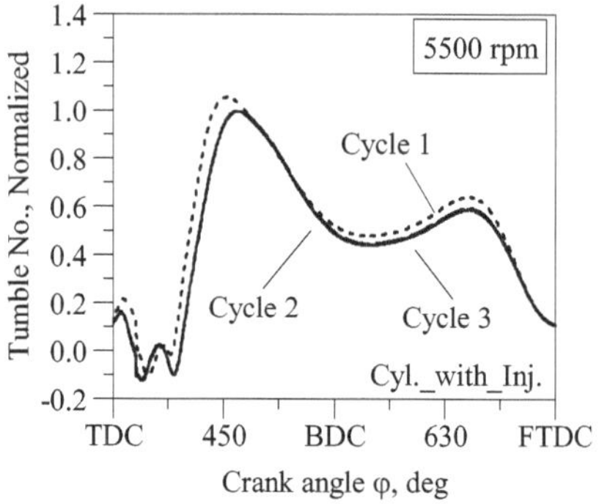

Figure 11.52: *Tumble ratio (-)*
- 5500 rpm WOT – Cylinder with injection -
Different simulation cycles.

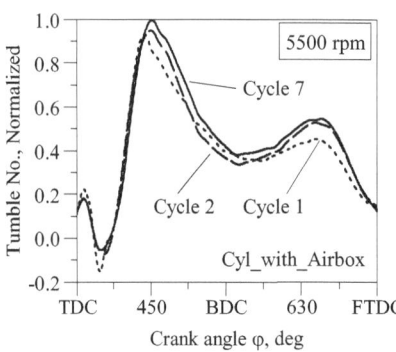

Figure 11.53: *Tumble ratio (-)*
- 5500 rpm WOT – Cylinder with airbox -
Different simulation cycles.

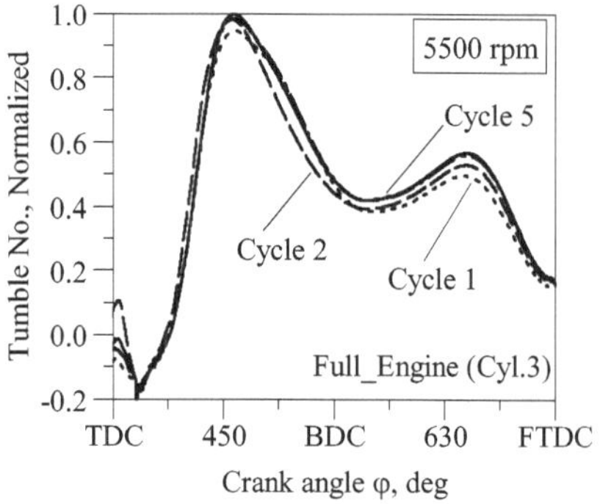

Figure 11.54: *Tumble ratio (-)*
- 5500 rpm WOT – Full engine (cyl.3) -
Different simulation cycles.

Focusing now on the results of cylinder 3 of the full engine simulation at 5500 rpm the results of a few relevant variables during the convergence process are reported. This helps understanding the transition from the "running-in phase" (see Chapters 7.2.1 and 11.3) to numerical cycle-to-cycle variations at the convergence. Depending on the phenomenon behind a variable the required convergence time is different. The mass flow through the intake valves, the trapped air mass in the cylinder and the turbulence (see Figures 11.55, 11.56 and 11.58), which can be considered representative for the motion of the fresh charge during the exchange period show a good convergence starting with the third simulated cycle (at low load operating conditions or in case of a bigger airbox it would surely take longer). From that point numerical cycle-to-cycle variations are still present and are intrinsic in the engine behavior.

Considering the convergence of the mixture composition (lambda value in Figure 11.57) it can be seen that a very good estimation of the initial conditions at this operating point (see Chapter 11.3) allows reaching the target value of fuel in the cylinder already in the second cycle. At other operating points the convergence of lambda is usually reached with the fourth cycle.

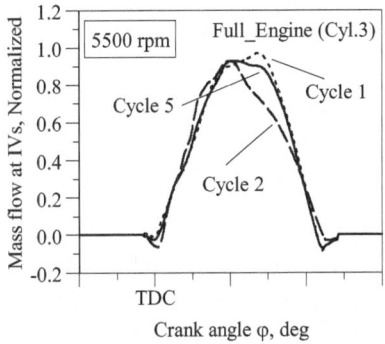

Figure 11.55: *Mass flow through IVs (mg/s) - 5500 rpm WOT – Full engine (cyl.3) - Different simulation cycles.*

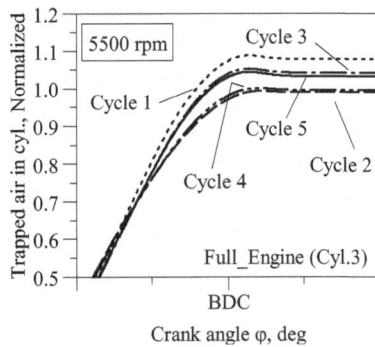

Figure 11.56: *Air mass trapped in cyl. (mg/s) - 5500 rpm WOT – Full engine (cyl.3) - Different simulation cycles.*

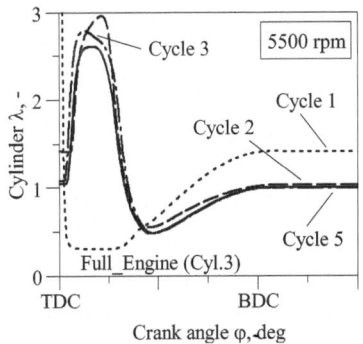

Figure 11.57: *Lambda in cyl. (mg/s) - 5500 rpm WOT – Full engine (cyl.3) - Different simulation cycles.*

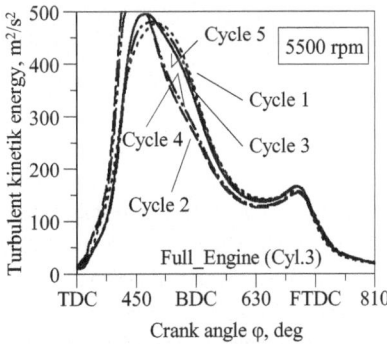

Figure 11.58: *Turbulence in cyl. (mg/s) - 5500 rpm WOT – Full engine (cyl.3) - Different simulation cycles.*

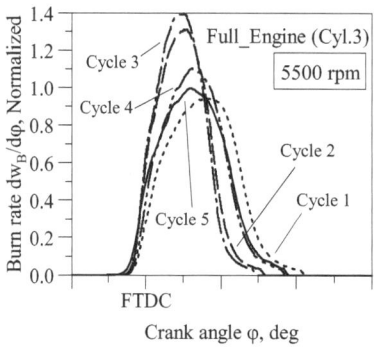

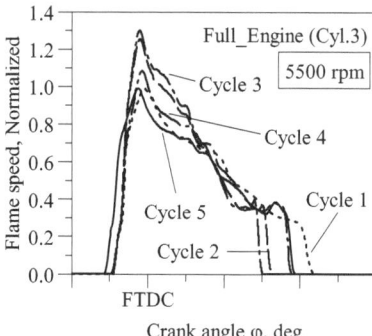

Figure 11.59: *Cylinder burn rate (%/deg) - 5500 rpm WOT – Full engine (cyl.3) - Different simulation cycles.*

Figure 11.60: *Abs. flame speed (m/s) - 5500 rpm WOT – Full engine (cyl.3) - Different simulation cycles.*

The last Figures 11.59 and 11.60 refer to the combustion process. Usually this process, especially in case of "critical" mixture formation like in this engine, requires more simulated cycles in order to bring the results to intrinsic cycle-to-cycle variations. Here it can be seen that the first three cycles cannot be taken as representative, not until the fourth cycle the results satisfiy the convergence criteria (in this simulation the results of 15 cycles have been investigated).

11.9 Result Comparison among the different Extensions of the 3D-CFD-Domain

Concluding, the comparisons of a few variables of cylinder 3 among the different extensions of the 3D-CFD-domain presented in this Chapter are reported. The analysis of the intake mass flow through the valves show remarkable differences among the cases (see Figure 11.61). The simulations without airbox have profiles affected by oscillations caused by the vicinity of the intake boundary condition and in particular in case of "Cyl_without_Inj." (see Chapter 11.5.1) the peak is at the highest. The profile of the intake mass flow between the two simulations with airbox is quite similar. In case of the simulation of one cylinder with the airbox, probably due to the influence of the 0D- and 1D-models for the simulation of the missing cylinders, the mass flow is shifted towards the valve opening phase. The mass fraction of residual gas at IVC shows small differences (see Figure 11.62). In particular the simulations with the airbox show the lowest concentration of residual gas.

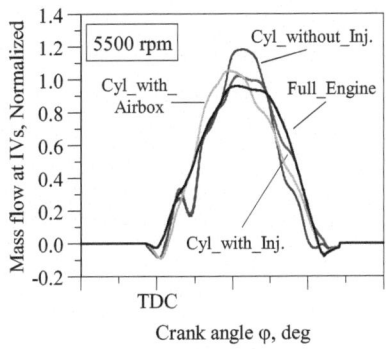

Figure 11.61: *Mass flow through IVs (mg/s) - 5500 rpm WOT – Different CFD-domains.*

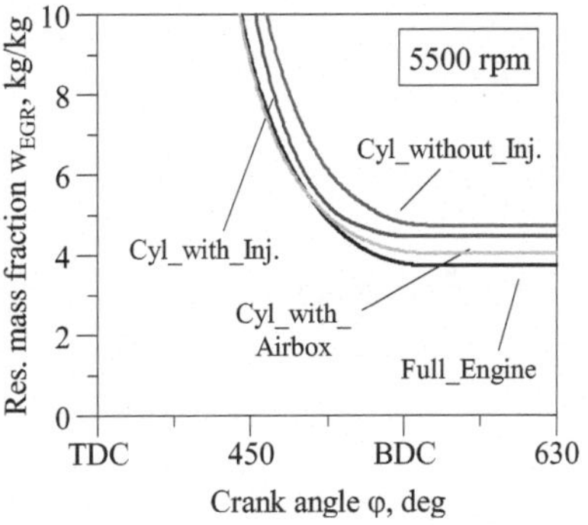

Figure 11.62: *Cylinder EGR mass fraction - 5500 rpm WOT – Different CFD-domains.*

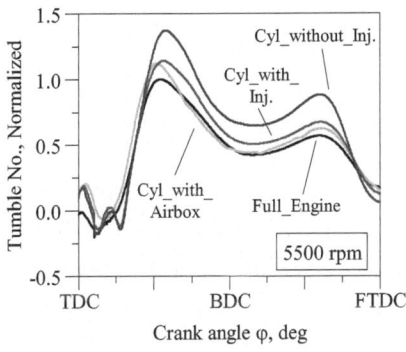

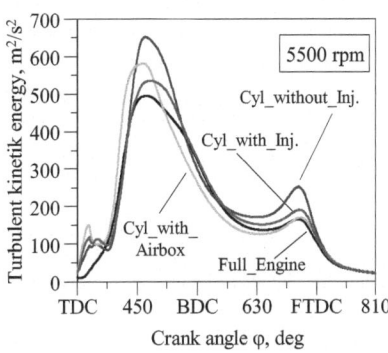

Figure 11.63: *Tumble ratio (-) - 5500 rpm WOT – Different CFD-domains.*

Figure 11.64: *Cylinder turbulence (TKE) - 5500 rpm WOT – Different CFD-domains.*

The differences among the tumble ratio evaluations are noticeable (see Figure 11.63). Here it can be seen that the nearer the intake 1D-CFD-boundary condition is to the valves, the higher is the tumble ratio. Reliable results are performed only including the airbox in the 3D-CFD-domains. Consequently, as usual in SI-engines with a considerable tumble level, the turbulence in the cylinder follows the trend of the tumble ratio (see Figure 11.64).

12

Conclusion

Since 1998, during my activity at the FKFS and IVK University of Stuttgart an innovative fast-response 3D-CFD tool has been developed. This tool aims for a better integration in the development process of internal combustion engines and represents a step towards virtual engine development. A more reliable calculation of the engine processes for engineering purposes and a clear representation of the results allow a more comprehensive analysis of the engine behavior. In addition comparisons with experimental data or results from other simulation tools can be easily and efficiently performed. The 3D-CFD-domain is not, as usually, limited to the simulation of the cylinder alone, but can be extended up to the full engine. This allows a more predictive simulation by reducing the effects of the boundary conditions on the results. The simulation run is always performed over many operating cycles so that the convergence of the results can also be proofed.

Within a 3D-CFD-simulation with *QuickSim* both an evaluation tool and a real working-process analysis are operative. This strategy allows an "online" analysis of the average relevant results from the flow field and establishes a comparative investigation with the real working-process analysis under the same thermodynamic conditions. This information network helps not only better investigating the phenomena occurring in the engine, but also sets the requisites for a more advanced and reliable modeling of engine processes. Each 3D-CFD-engine-model is a modular object within a network of local and global variables, which in dependence of their numerical reliability can be easily used and combined in the chosen formulation ("internal coupling"). Following this procedure the local implementation of 3D-CFD-engine-models, e.g., can also rely on typical and reliable phenomenological correlations used for global investigations of engine processes. Concluding, the engine-process modeling in *QuickSim* aims pragmatically to find the best compromise among the physical understanding of the phenomenon, the mathematical formulation, the algebraic description and the numerical implementation of the model also using coarse meshes. In contrast to the common practice, the dimensions of the cells within the 3D-

CFD-meshes, their structure and orientation are explicitly taken into account so that the reliability of the results in combination with an impressive reduction of the computational time (up to a factor 100) can be ensured.

Thanks to the high modularity of the program, different engine models can be easily tested and validated. This development procedure, which is a continuous process, improves with increasing of expertise, measurement capability, phenomena understanding, performance of numerical algorithms, etc. Following the development trend of simulation tools and considering the evolution of computational capabilities and their relatively low impact on the business plan for engine design, a more significant role of simulation tools in the development of future IC-engines is obvious.

13

Outlook

The fast response 3D-CFD-code *QuickSim* is a continuously improving tool with the aim to ensure reliability, versatility and efficiency in an increasing number of applications in the engine development process. In addition to the improvement of the already implemented 3D-CFD-models, the focus for the future development of *QuickSim* will be mainly on the following topics:

- Modeling of bio-fuels that may have a more complex composition than traditional hydrocarbons $C_nH_mO_rN_q$. In addition, for this kind of fuels, an efficient determination of the laminar flame speed based on pressure trace measurements at the test bench will be investigated.

- Simulation of dual-fuel engines (contemporaneous injection of two different fuels, e.g. diesel direct injection combined with natural gas manifold injection).

- Modeling of HCCI engines with improved self-ignition models and a reliable numerical control in the switching step between HCCI and traditional operation condition.

- Exhaust emission models based on a local implementation of phenomenological approaches using an "internal coupling" with the real working-process analysis.

- Multi processor solution with an efficient allocation of each cylinder to a processor. The interfaces among the 3D-CFD-domains are located in the manifolds where the number of shared cells is limited. Therefore the computation time can be supplementary reduced without remarkable overheads in the information exchange process among the different 3D-CFD-domains.

- A reliable coupling with external 1D-CFD-programs e.g. for the simulation of turbo-charging and the rest of the intake and exhaust systems.

- An improved integration with other simulation programs. Exemplary *QuickSim* could act as a calibration device for the real working-process analysis in case of virtual engines where experimental data from the test bench are not available. In additional investigations towards a better comparability of results under different environmental conditions will be performed.

- Automatic engine design modifications, in particular piston shape and intake channels, towards an increasing of the engine efficiency.

All these topics will be the result of intense work in the future. Future publications will report the relevant evolution steps of the simulation tool proposed in this work.

Appendix A

A.1 Vector and Matrix Analysis

Some definitions and laws from the vector- and matrix-analysis, used in this work are explained in details here. S denotes a scalar, $\vec{v}$ a vector and $\overline{\overline{T}}$ a matrix (second-order tensor):

$$\vec{v} = \begin{pmatrix} v_x \\ v_y \\ v_z \end{pmatrix} \qquad \overline{\overline{T}} = \begin{pmatrix} T_{xx} & T_{xy} & T_{xz} \\ T_{yx} & T_{yy} & T_{yz} \\ T_{zx} & T_{zy} & T_{zz} \end{pmatrix}.$$

The *dyadic product* $\vec{v} \otimes \vec{v}'$ of two vectors $\vec{v}$ and $\vec{v}'$ leads to a matrix $\overline{\overline{T}}$:

$$\vec{v} \otimes \vec{v}' = \begin{pmatrix} v_x v'_x & v_x v'_y & v_x v'_z \\ v_y v'_x & v_y v'_y & v_y v'_z \\ v_z v'_x & v_z v'_y & v_z v'_z \end{pmatrix}.$$

The *transposed matrix* $\overline{\overline{T}}^T$ is obtained by interchanging rows and columns of $\overline{\overline{T}}$:

$$\overline{\overline{T}}^T = \begin{pmatrix} T_{xx} & T_{yx} & T_{zx} \\ T_{xy} & T_{yy} & T_{zy} \\ T_{xz} & T_{yz} & T_{zz} \end{pmatrix}.$$

Furthermore, the *unit matrix* $\overline{\overline{I}}$ is defined as:

$$\overline{\overline{I}} = \begin{pmatrix} 1 & 0 & 0 \\ 0 & 1 & 0 \\ 0 & 0 & 1 \end{pmatrix}.$$

The *contraction* $\overline{\overline{T}} : \overline{\overline{T}}'$ of two matrices $\overline{\overline{T}}$ and $\overline{\overline{T}}'$ yields a scalar S by means of:

$$\overline{\overline{T}} : \overline{\overline{T}}' = \sum_i \sum_j T_{ij} \cdot T'_{ji} = S.$$

The *gradient* of a scalar S yields a vector of the form:

$$grad\ S = \begin{pmatrix} \dfrac{\partial S}{\partial x} \\[2mm] \dfrac{\partial S}{\partial y} \\[2mm] \dfrac{\partial S}{\partial z} \end{pmatrix}.$$

The *gradient* of a vector $\bar{v}$ yields a matrix of the form:

$$grad\ \bar{v} = \begin{pmatrix} \dfrac{\partial v_x}{\partial x} & \dfrac{\partial v_y}{\partial x} & \dfrac{\partial v_z}{\partial x} \\[2mm] \dfrac{\partial v_x}{\partial y} & \dfrac{\partial v_y}{\partial y} & \dfrac{\partial v_z}{\partial y} \\[2mm] \dfrac{\partial v_x}{\partial z} & \dfrac{\partial v_y}{\partial z} & \dfrac{\partial v_z}{\partial z} \end{pmatrix}.$$

The *divergence* of a vector $\bar{v}$ yields a scalar:

$$div\ \bar{v} = \frac{\partial v_x}{\partial x} + \frac{\partial v_y}{\partial y} + \frac{\partial v_z}{\partial z}.$$

The *divergence* of a matrix $\overline{\overline{T}}$ yields a vector:

$$div\ \overline{\overline{T}} = \begin{pmatrix} \dfrac{\partial T_{xx}}{\partial x} + \dfrac{\partial T_{yx}}{\partial y} + \dfrac{\partial T_{zx}}{\partial z} \\[2mm] \dfrac{\partial T_{xy}}{\partial x} + \dfrac{\partial T_{yy}}{\partial y} + \dfrac{\partial T_{zy}}{\partial z} \\[2mm] \dfrac{\partial T_{xz}}{\partial x} + \dfrac{\partial T_{yz}}{\partial y} + \dfrac{\partial T_{zz}}{\partial z} \end{pmatrix}.$$

Appendix B

B.1 Thermodynamic Properties of the Working Fluid

In this appendix the thermodynamic properties of the burned gas of a CNG-engine are reported (see Chapters 8 and 11).

Fuel Specifications of Natural Gas

Natural Gas: $C_{1.02}H_{4.02}N_{0.01}$ with LHV = 49.14 MJ/kg.

Composition of Burned Gas after a Combustion of Natural Gas

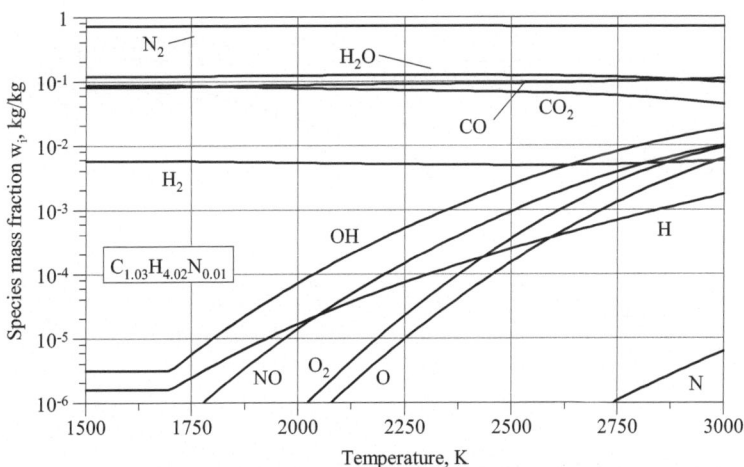

Figure B.1: *Burned gas composition as function of temperature - λ =0.7 and p =1bar (fuel: natural gas $C_{1.03}H_{4.02}N_{0.01}$).*

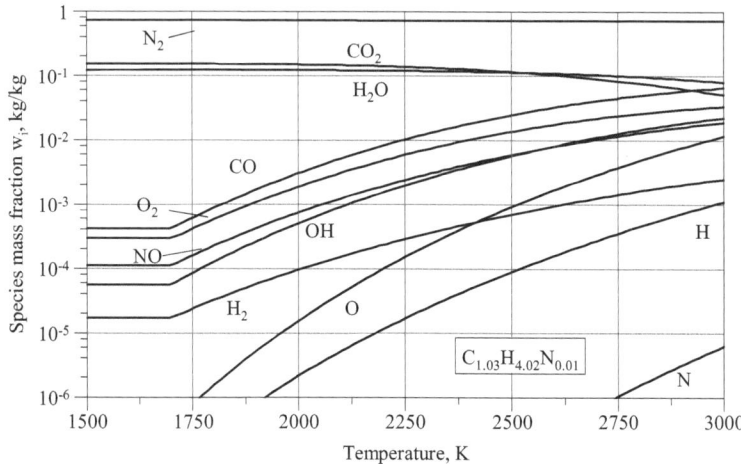

Figure B.2: *Burned gas composition as function of temperature -* $\lambda = 1.0$ *and* $p = 1bar$
(fuel: natural gas $C_{1.03}H_{4.02}N_{0.01}$).

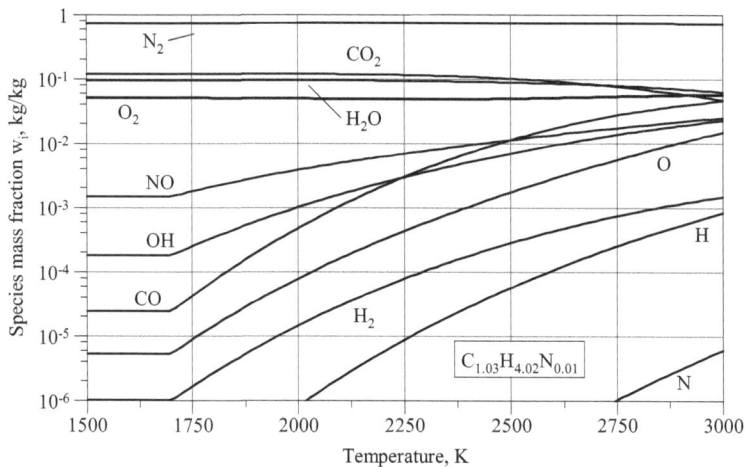

Figure B.3: *Burned gas composition as function of temperature -* $\lambda = 1.3$ *and* $p = 1bar$
(fuel: natural gas $C_{1.03}H_{4.02}N_{0.01}$).

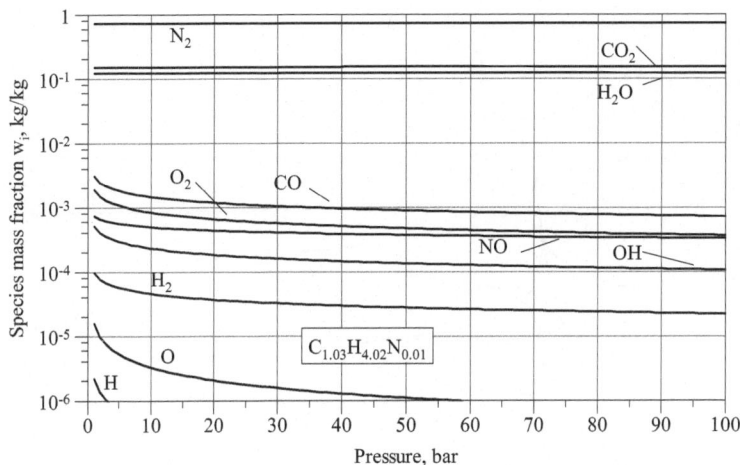

Figure B.4: *Burned gas composition as function of pressure -* $\lambda = 1.0$ *and* $T = 2000K$ *(fuel: natural gas $C_{1.03}H_{4.02}N_{0.01}$).*

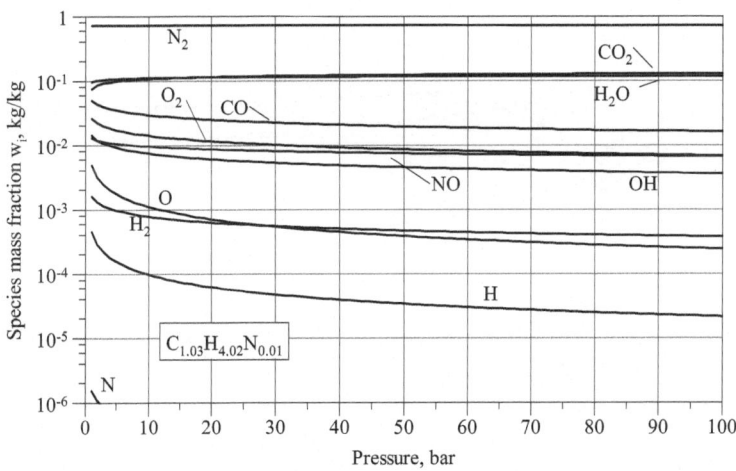

Figure B.5: *Burned gas composition as function of pressure -* $\lambda = 1.0$ *and* $T = 2800K$ *(fuel: natural gas $C_{1.03}H_{4.02}N_{0.01}$).*

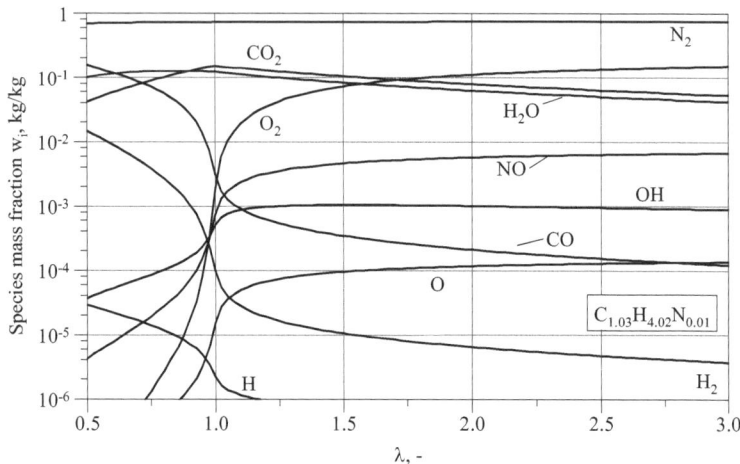

Figure B.6: *Burned gas composition as function of lambda - p =1bar and T =2000K (fuel: natural gas $C_{1.03}H_{4.02}N_{0.01}$).*

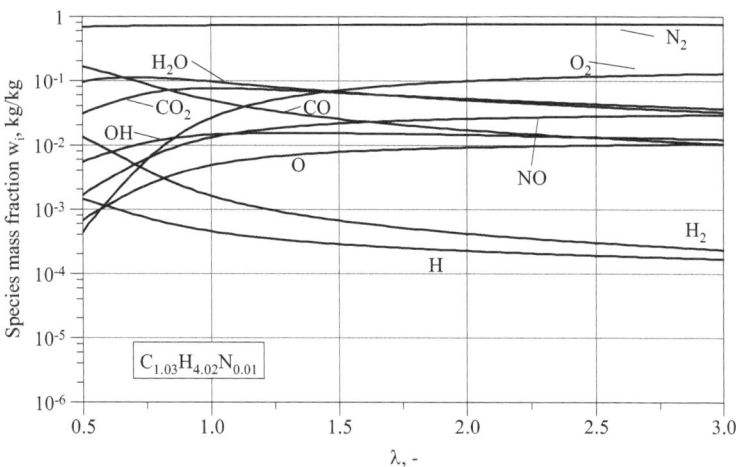

Figure B.7: *Burned gas composition as function of lambda - p =1bar and T =2800K (fuel: natural gas $C_{1.03}H_{4.02}N_{0.01}$).*

Thermal Properties of Burned Gas after a Combustion of Natural Gas

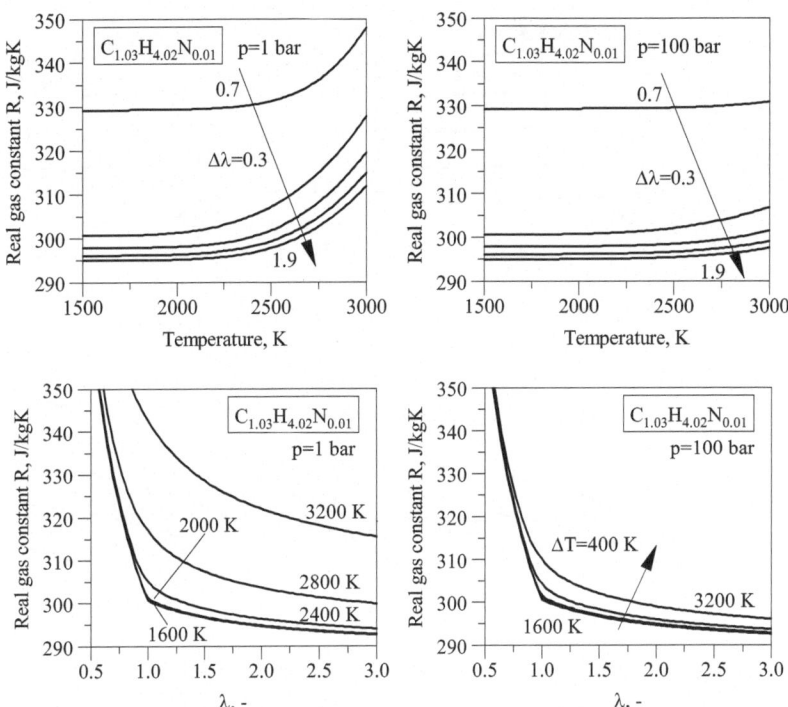

Figure B.8: *The real gas constant R as function of temperature and lambda.*

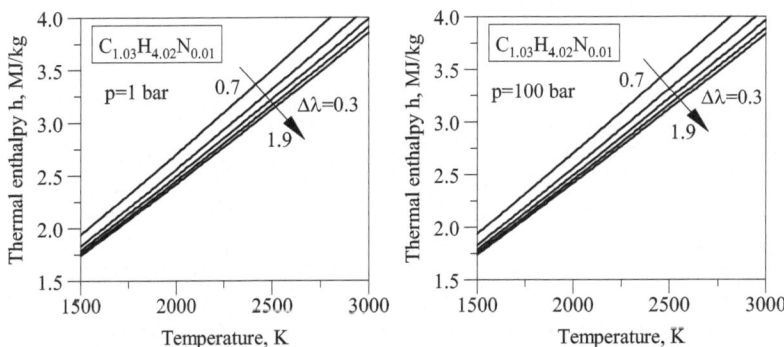

Figure B.9: *The thermal enthalpy h as function of temperature.*

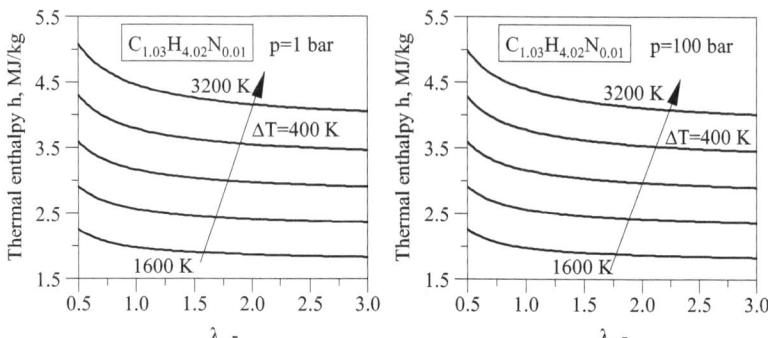

Figure B.10: *The thermal enthalpy h as function of lambda.*

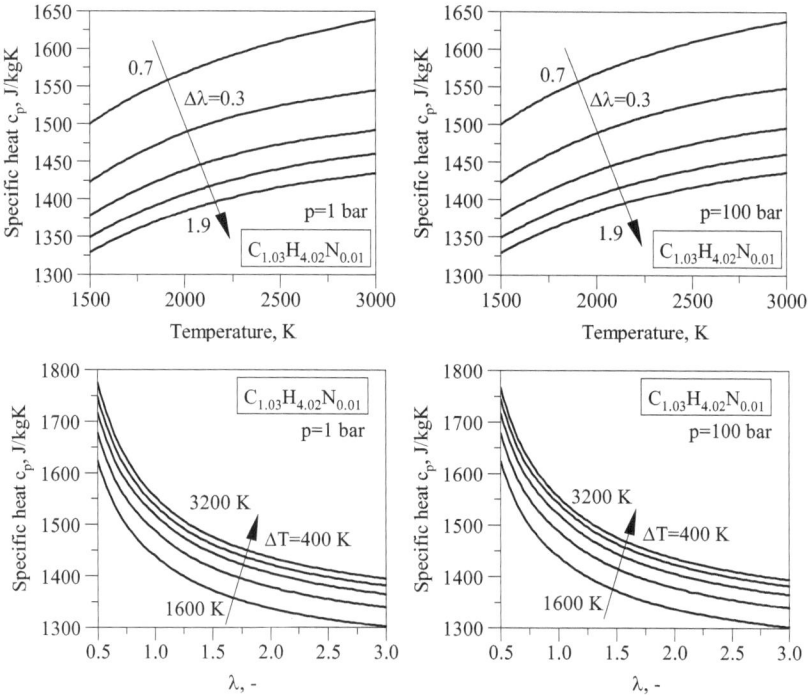

Figure B.11: *The specific heat c_p as function of temperature and lambda.*

Enthalpy of Formation of Burned Gas after Combustion with Natural Gas

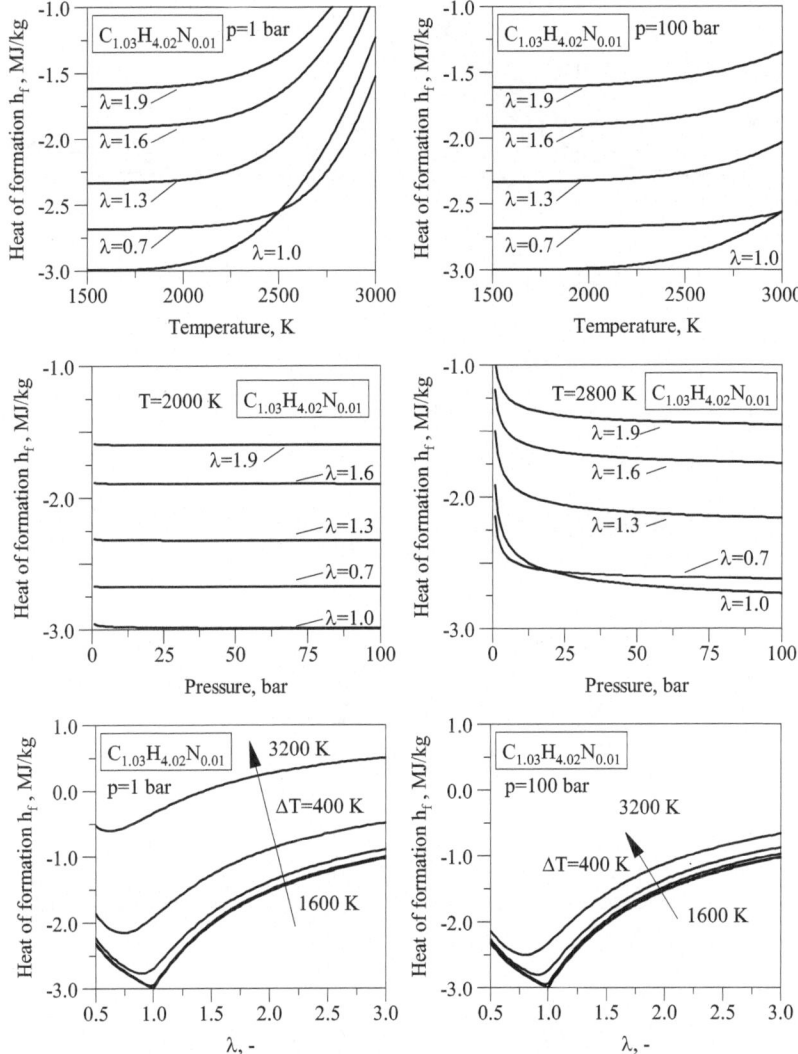

Figure B.12: *The enthalpy of formation h_f as function of temperature, pressure and lambda.*

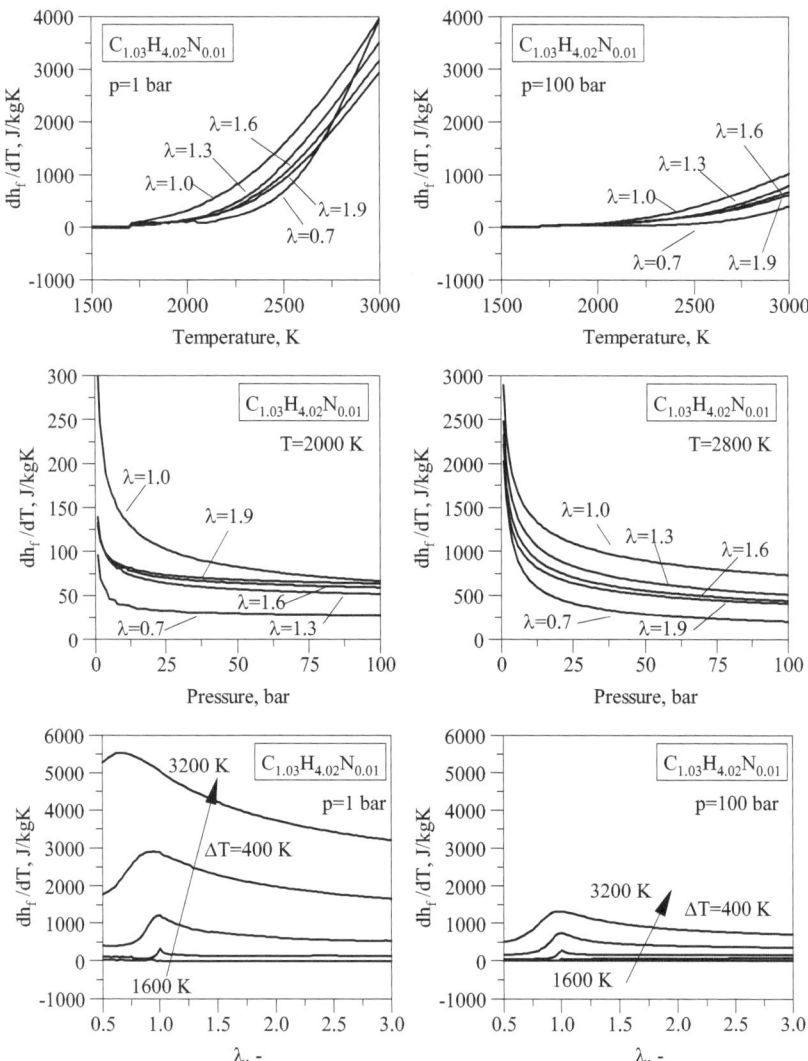

Figure B.13: *Gradient of heat of formation: dh_f / dT*
as function of temperature, pressure and lambda.

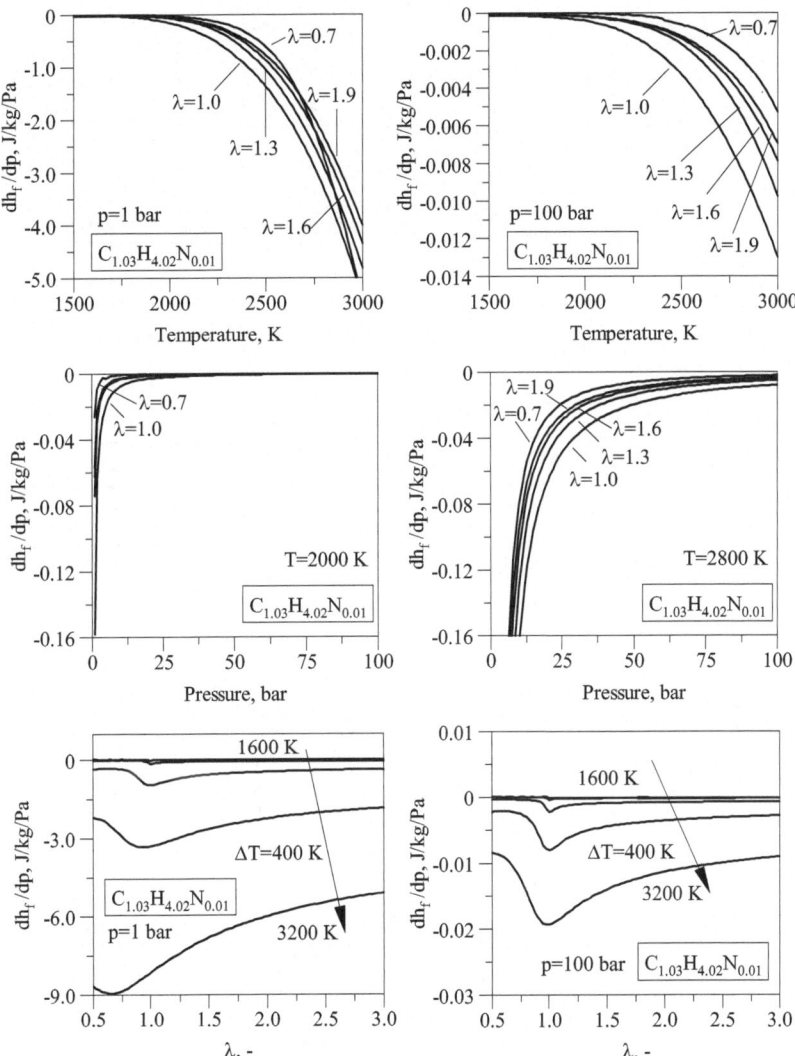

Figure B.14: *Gradient of heat of formation: dh_f/dp*
as function of temperature, pressure and lambda.

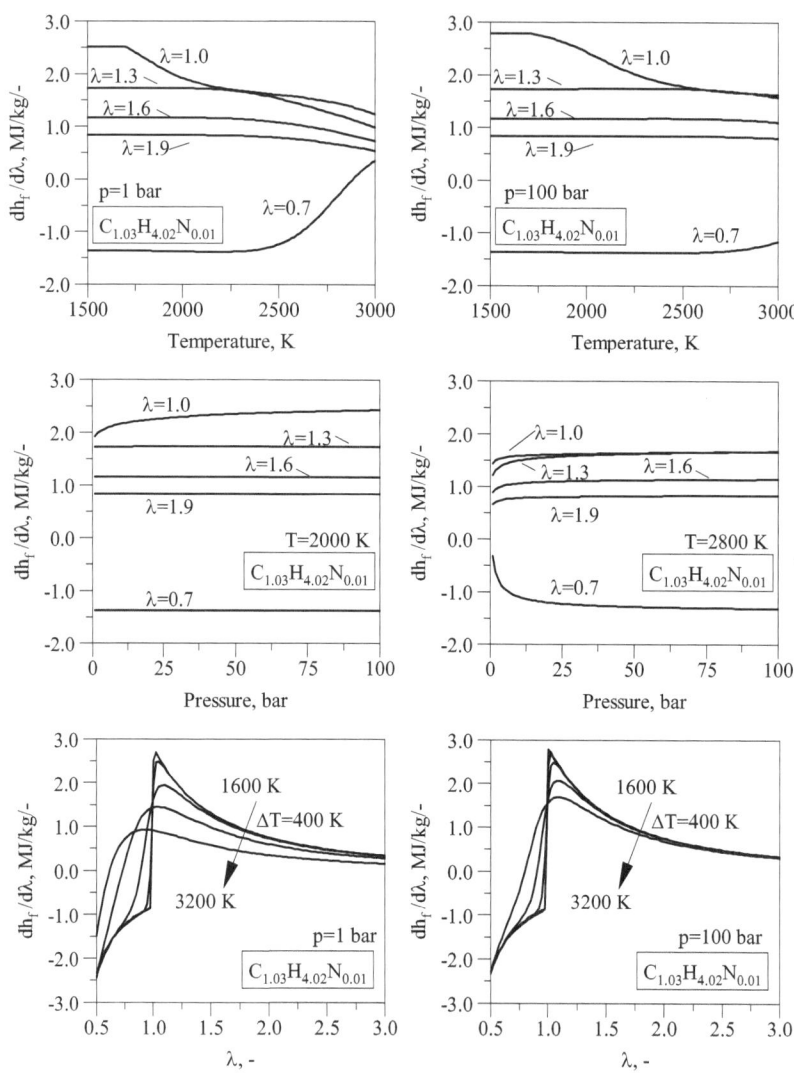

Figure B.15: *Gradient of heat of formation: $dh_f/d\lambda$ as function of temperature, pressure and lambda.*

Combustion Conversion Efficiency of Natural Gas

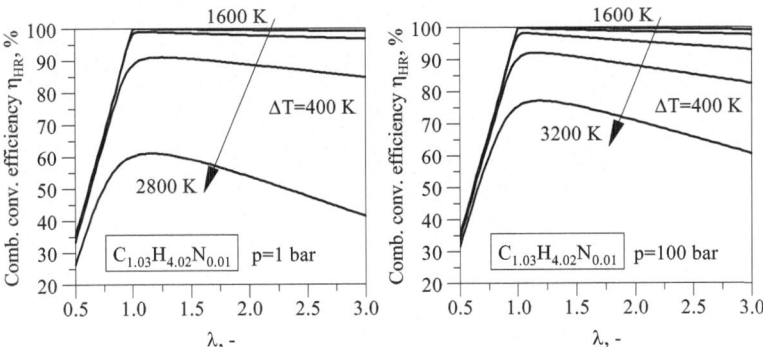

Figure B.16: *Combustion conversion efficiency* η_{HR} *(p =1 bar and p =100 bar).*

References

[1] WIKIPEDIA - The free encyclopedia: *Ford Model T*

[2] MELOSI M.: *The Automobile and the Environment in American History - Auto Emissions and Air Pollution.* www.autolife.umd.umich.edu/Environment/ E_Overview/E_Overview1.htm

[3] GUEZENNEC Y., MIDLAM-MOHLER S., RIZZONI G., HAAS S., BERNER H.-J., BARGENDE M.: *Mixed-Mode Diesel HCCI/DI with External Mixture Preparation.* FISITA 2004, Barcelona, Paper F2004V258. Article

[4] WIKIPEDIA - The free encyclopedia: *Bio-fuels*

[5] HEYWOOD J.-B.: *Internal Combustion Engine Fundamentals.* McGraw-Hill International Editions, New York, 1988.

[6] BARGENDE M.: *Verbrennungsmotoren I,II und III.* Lecture Manuscript, Universität Stuttgart, 2009.

[7] BARGENDE M.: *Berechnung und Analyse innermotorischer Vorgänge bei Verbrennungsmotoren.* Lecture Manuscript, Universität Stuttgart, 2009.

[8] GROHE H. : *Otto- und Dieselmotoren.* 11th ed., Vogel Publisher, Würzburg, 1995.

[9] SASS F.: *Geschichte des Deutschen Verbrennungsmotorenbaues, von 1860 bis 1918.* Springer Publisher, Berlin, 1962.

[10] BARGENDE M., RIEGLER U., SCHOLZ B.: *Der virtuelle Motor – Fiktion oder Realität ?.* Tagung Im Haus der Technik, Essen, October 2000.

[11] OTTO F., DITTRICH P., WIRBELEIT F., *Status of 3D-Simulation of Diesel Combustion.* 3. Internationales Indiziersymposium, 1998.

[12] GRILL M.: *Objektorientierte Prozessrechnung von Verbrennungsmotoren.* Dissertation, Universität Stuttgart, expert Verlag 2006.

[13] SCHMID A., GRILL M., BERNER H.-J., BARGENDE M., ROSSA S., BÖTTCHER M.: *Development of a Quasi-Dimensional Combustion Model for Stratified SI Engines.* SAE Paper 2009-01-2659.

[14]   GRILL M., BARGENDE M., SCHMID A., RETHER D.: *Quasi-Dimensional and Empirical Modeling of Compression-Ignition Engine Combustion and Emissions.* SAE Paper 2010-01-0151.

[15]   RETHER D., SCHMID A.,GRILL M., BARGENDE M.: *Quasi-Dimensional Modeling of CI-Combustion with Multiple Pilot- and Post Injections.* SAE Paper 2010-01-0150.

[16]   GRILL M., BARGENDE M.: *The Development of an Highly Modular Designed Zero-Dimensional Engine Process Calculation Code.* SAE Paper 2010-01-0149.

[17]   GRILL M., BARGENDE M.: *Das Zylindermodul: Neue Simulation nicht nur für zukünftige Brennverfahren.* MTZ 10/2009.

[18]   GRILL M., BILLINGER T., BARGENDE M.: *Quasi-Dimensional Modeling of Spark Ignition Engine Combustion with Variable Valve Train.* SAE Paper 2006-01-1107.

[19]   URLAUB A.: *Verbrennungsmotoren.* 2nd Edition, Springer Publisher, Berlin, Heidelberg, New York, 1994.

[20]   FUCHS H.-U.: The Dynamics of Heat. *Springer Publisher.* 1995.

[21]   CHIODI M., BARGENDE M.: *Improvement of Engine Heat-Transfer Calculation in the Three-Dimensional Simulation Using a Phenomenological Heat-Transfer Model.* SAE-Paper 2001-01-3601.

[22]   CHIODI M., BARGENDE M.: *Kopplung von 3D-CFD-Simulation und Arbeits-prozessrechnung als Bestandteil einer genaueren drei-dimensionalen Wand-wärmeübergangsrechnung in Verbrennungsmotoren.* 8. Tagung: Der Arbeitsprozess des Verbrennungsmotors. Graz, September 2001.

[23]   BADAMI M., BEVILACQUA V., MILLO F., CHIODI M., BARGENDE M.: *GDI Swirl Injector Spray Simulation: a Combined Phenomenological-CFD Approach.* SAE Paper 2004-01-3005.

[24]   PISCHINGER R., KRAßNIG G., TAUCAR G., SAMS T.: *Thermodynamik der Verbrennungsmaschine.* Springer Publisher, Wien, New York, 1993.

[25]   PFLAUM W.,MOLLENHAUER K.: *Wärmeübergang in der Verbrennungskraftmaschine.* Springer-Verlag, Wien, (1977).

[26]   BARGENDE M.: *Ein Gleichungsansatz zur Berechnung der instationären Wandwärmeverluste im Hochdruckteil von Ottomotoren.* Dissertation, Technische Universität Darmstadt, 1991.

[27]   JUSTI E.: *Spezifische Wärme, Enthalpie, Entropie und Dissoziation technischer Gase.* Springer Publisher, Berlin 1938.

[28]   ZACHARIAS F.: *Analytische Darstellung der thermodynamischen Eigenschaften von Verbrennungsgasen.* Dissertation, Technische Universität Berlin, 1966.

[29]   GORDON S., MCBRIDE B.-J.: *Computer Program for the Calculation of Complex Chemical Equilibrium Composition, Rocket Performance, Incident and Reflected Shocks, and Chapman-Jouguet Detonations.* NASA Publication SP-273, 1971.

[30]   DE JAEGHER P.: *Einfluss der Stoffeigenschaften der Verbrennungsgase auf die Motorprozessrechnung.* Habilitation, Technische Universität Graz, 1984.

[31]   HRUSTANOVIC S.: *Entwicklung eines Neuronalen Netzes zur schellen Berechnung der kalorischen Eigenschaften des Arbeitsmittels in Verbrennungsmotoren.* Diplomarbeit, IVK - Universität Stuttgart, 2002.

[32]   REINHARDT S.: *Entwicklung eines Neuronalen Netzes zur besseren Berechnung der Kalorik in CNG-Motoren.* Diplomarbeit, IVK - Universität Stuttgart, 2003.

[33]   BERNER H.-J., CHIODI M., BARGENDE M.: *Berücksichtigung der Kalorik des Kraftstoffes Erdgas in der Prozessrechung.* 9. Tagung: Der Arbeitsprozess des Verbrennungsmotors. Graz, September 2003.

[34]   GRILL M., SCHMID A., CHIODI M., BERNER H.-J., BARGENDE M.: *Calculating the Properties of User-Defined Working Fluids for Real Working-Process Simulations.* SAE Paper 2007-01-0936.

[35]   GRILL M., CHIODI M., BERNER H.-J., BARGENDE M.: *Berechnung der thermodynamischen Stoffwerte von Rauchgas und Kraftstoffdampf beliebiger Kraftstoffe.* MTZ 05/2007.

[36]   PRANDTL L., OSWATITSCH K., WIEGHARDT K.: *Führer durch die Strömungslehre.* 9th Edition, Vieweg Editions, Braunschweig, 1979.

[37]   SCHLICHTING H.: *Boundary Layer Theory.* 7th Edition, McGraw-Hill International Editions, New York, 1988.

[38]   WILLUMEIT H.-P.: *Der Wärmeübergang im Verbrennungsmotor.* MTZ 47, (1986), 1

[39]   WOSCHNI G.: *Universal Applicable Equation for the Instantaneous Heat transfer Coefficient in the Internal Combustion Engine.* SAE Paper 670931.

[40]   HOHENBERG G.-F.: *Advanced Approaches for Heat Transfer Calculations.* SAE Paper 790825.

[41]   WIMMER A., PIVEC R.: *Experimental and Numerical Investigations on Gas-Side Heat Transfer.* 7. Tagung: Der Arbeitsprozess des Verbrennungsmotors. Graz, September 1999.

[42]   VOGT R.: *Beitrag zur rechnerischen Erfassung der Stickoxidbildung in Dieselmotoren.* Dissertation, Universität Stuttgart 1975.

[43]   WIEBE I.-I.: *Brennverlauf und Kreisprozess von Verbrennungsmotoren.* Berlin: Verl. Technik, 1970.

[44]   WICHELHAUS D.: *Optimierung der dieselmotorischen Verbrennung für hohe Leistung bei begrenztem Luftangebot.* Dissertation, Universität Stuttgart, expert Verlag 2009.

[45]   BARBA C., BOULOUCHOS K., BARGENDE M.: *Empirische und phänomenologische Verbrennungsmodelle beim schnelllaufenden CR-Dieselmotor. Ergänzung zu Versuch und CFD-Simulation.* 4rd Stuttgarter Symposium, 2001.

[46]   GÜLDER OE.-L.: *Correlations of Laminar Combustion Data for Alternative S.I. Engine Fuels.* SAE-Paper 841000.

[47]   WINTERBONE D.-E., PEARSON R.-J.: *Design Techniques for Engine Manifolds.* Professional Engineering Publishing, London, 1999.

[48]   N. N.: *GT-POWER User's Manual and Tutorials.* Gamma Technologies, Chicago, 2007.

[49]   SCROCCO P.: *1D Simulation for the optimization of the intake and exhaust systems of a race engine.* Diplomarbeit IVK – Universität Stuttgart and Politecnico di Torino, 2008.

[50]   GIORDANINO A.: *Optimization of a Formula 3 engine for different environmental conditions.* Diplomarbeit IVK – Universität Stuttgart and Politecnico di Torino, 2008.

[51]   SERVETTO E.: *Optimization of a high performance diesel engine by means of one-dimensional numerical simulation.* Diplomarbeit IVK – Universität Stuttgart and Politecnico di Torino, 2009.

[52]   CHIODI M.: *Dreidimensionale Simulation der Strömung in einem Ottomotor während des Ladungswechsels.* Diplomarbeit IVK - Universität Stuttgart and Politecnico di Milano, 1998.

[53]   NOLL B.: *Numerische Strömungsmechanik.* Springer Publisher, Berlin, Heidelberg, New York, 1993.

[54]   ANDERSON J.-D. JR.: *Computational Fluid Dynamics.* McGraw-Hill International Editions, New York, 1995.

[55]   WARNATZ J., MAAS U.: *Combustion.* 3rd Edition, Springer Publisher, Berlin, Heidelberg, New York, 1993.

[56]   N. N.: *Star-CD User Manual Version 3.15.* Adapco (Computational Dynamics Ltd.). London, 2001.

[57]   N.N.: *JANAF Thermochemical Tables.* USA, National Bureau of Standards, 1971.

[58]   HIRSCHFELDER J., CURTIS C., BIRD R.: *Molecular theory of gases and liquids.* Wiley, 1964.

[59]   LIBBY P.-A., WILLIAMS F.-A.: *Turbulent Reacting Flows. Academic Press,* 1993.

[60]   BEVILACQUA V.: *3D-CFD Analysis and Modeling of Fuel-Spray-Injection for GDI-Engines.* Diplomarbeit IVK – Universität Stuttgart and Politecnico di Torino, 2003.

[61]   RIEGLER U.-R.: *Berechnung der Verbrennung und der Schadstoffbildung in Ottomotoren unter der Verwendung detaillierter Reaktionsmechanismen.* Dissertation, Universität Stuttgart 1999.

[62]   PAULE K., WICHELHAUS D., MACK O., CHIODI M., BARGENDE M.: *Investigation on different Injection Concepts of a High-Performance Diesel Engine by means of the 3D-CFD-Development-Tool QuickSim.* 9th Stuttgarter Symposium, March 2009.

[63]   CHIODI M., MACK O., BARGENDE M., PAULE K., BRANDT V. FACKH J., WICHELHAUS D.: *Improvement of a High-Performance Diesel-Engine by means of Investigation on different Injection Strategies.* SAE Paper 2009-24-0008.

[64]   Paule K., Wichelhaus D., Ferrari A., Chiodi M., Bargende M.: CNG im Rennsport. Gasfahrzeugekonferenz, October 2009.

[65]   GUSMERINI L.: *Investigation on the Gas Exchange Process in Four-Strokes Spark Ignition Engines.* Diplomarbeit IVK – Univ. Stuttgart and Politecnico di Torino, 2002.

[66]   ZUARINI A.: *Numerical Analysis of Combustion Process in Four-Strokes Spark Ignition Engines.* Diplomarbeit IVK – Universität Stuttgart and Politecnico di Torino, 2002.

[67]   ZÜLCH C., KULZER A., CHIODI M., BARGENDE M.: *Directstart: Investigation of Mixture Formation by Means of Optical Measurements and 3D-CFD-Simulation.* SAE Paper 2005-01-3686.

[68]   ZÜLCH C.: *Konzepte für einen sicheren Direktstart von Ottomotoren.* Dissertation, Universität Stuttgart, Shaker Verlag 2007.

[69]   WELLER H.-G., USLU S., GOSMAN A.-D., MALY R.-R., HERWEG R., HEEL B.: *Prediction of Combustion in Homogeneous-Charge Spark-Ignition Engines.* International Symposium COMODIA 94.

[70]   HERWEG R., MALY R.-R.: *A Fundamental Model for Flame Kernel Formation in S.I. Engines.* SAE-Paper 922243.

[71]   SARGENTI R.: *Numerische Ermittlung von Brennraumwandtemperaturen bei Verbrennungsmotoren.* Dissertation, Universität Stuttgart 2006.

[72]   BOHAC S.-V., BAKER D.-M., ASSANIS D.-N.: *A Global Model for Steady-State and Transient S.I. Engine Heat Transfer Studies.* SAE Paper 960073, (SP-1168).

[73]   SHAYLER P.-J., MA T.: *A Model for the Investigation of Temperature, Heat Flow and Friction Characteristics During Engine Warm-Up.* SAE Paper 931153.

[74]   CHOI G.-H., CHOI K.-H.: *Analysis of Combustion Chamber Temperature and Heat Flux in a DOHC Engine.* SAE Paper 970895, (SP-1246).

[75]   LIMBACH S.: *Multi-Dimensional Computation of Transient Convective Heat Transfer: Application to a Reciprocating Engine.* Dissertation, TU Graz, 1997.

[76]   WOSCHNI G.: *Experimentelle Bestimmung des örtlich gemittelten Wärmeübergangs-Koeffizienten im Ottomotor.* MTZ 42, 1981, 6

[77]   HUBER K.: *Der Wärmeübergang schnelllaufender, direkteinspritzender Dieselmotoren.* Dissertation, Technische Universität München, 1991.

[78]   ZAPF H.: *Untersuchungen zur Vorausberechnung der Ladungsendtemperatur in Viertakt-Dieselmotoren.* MTZ, August 1970.

[79]   ZAPF H.: *Beitrag zur Untersuchung des Wärmeübergangs während des Ladungswechsels im Viertakt-Dieselmotor.* MTZ, December 1969.

[80]   JACKSON N.-S., PILLEY A.-D.: *Instantaneous Heat Transfer in a Highly Rated DI Truck Engine.* SAE Paper 900692, 1990.

[81]   JENNINGS M.-J., MOREL T.: *A Computational Study of Wall Temperature Effects on Engine Heat Transfer.* SAE Paper 910459.

[82]   MOREL T.: *An Improved Near-Wall Heat Transfer Model for Multidimensional Engine Flow Calculations.* SAE Paper 900251.

[83]   MOREL T., RACKMIL C.-I.: *Model for Heat Transfer and Combustion in Spark Ignited Engines and its Comparison with Experiments.* SAE Paper 880198.

[84]   REITZ R.-D.: *Assessment of Wall Heat Transfer Models for Premixed-Charge Engine Combustion Computations.* SAE Paper 910267.

[85]   SHAYLER P.-J., MAY S.-A.: *Heat Transfer to the Combustion Chamber Walls in Spark-Ignition Engine.* SAE Paper 950686.

[86]   WOLFF A., BOULOUCHOS K., MÜLLER R.: *A Computational Investigation of Unsteady Heat Flux Through an I.C. Engine Cylinder Wall Including Soot Layer Dynamics.* SAE Paper 970063, 1997.

[87]   VOLKSWAGEN MOTORSPORT: www.volkswagen-motorsport.com (24h Nürburgring).

[88]   CHIODI M., BERNER H.-J., BARGENDE M.: *Investigation on Mixture Formation and Combustion Process in a CNG-Engine by Using a Fast Response 3D-CFD-Simulation.* SAE Paper 2004-01-3004.

[89]   CHIODI M., BERNER H.-J., BARGENDE M.: *Investigation on Different Injection Strategies in a Direct-Injected Turbocharged CNG-Engine.* SAE Paper 2006-01-3000.

[90]   CHIODI M., BERNER H.-J., BARGENDE M.: *Schnelle 3D-CFD-Untersuchungen des Ansaugsystems eines aufgeladenen 1l-Erdgasmotors mit dem Entwicklungstool „QuickSim".* Tagung: Ansaugsysteme aufgeladener Motoren, Haus der Technik, Ludwigsburg, May 2006.

[91]   BOHATSCH S., HOFMANN B., FERRARI A., CHIODI M., BERNER H.-J., BARGENDE M.: *Darstellung verschiedener Betriebsstrategien an einem Erdgasmotor mit innerer Gemischbildung.* Tagung: Gasfahrzeuge, Haus der Technik, Dresden, July 2006.

[92]   FRIEDRICH C., QUARG J., BERNER H.-J., BARGENDE M.: *Leistungs- und Emissionspotenzial eines aufgeladenen 1 Liter-Erdgas-Motors.* 4. Stuttgarter Symposium, 2001.

[93]   ALLGEIER T., KLENK M., LANDENFELD T., FRIEDRIDCH C., QUARG J.: *Motorsteuerung für Erdgasfahrzeuge.* 3. Symposium „Steuerungssyteme für den Antriebsstrang von Kraftfahrzeugen", Berlin, 2001.

[94]   GEORGI B., MEHNERT J., DINGEL O.: *Möglichkeiten für CNG-Antriebe im Rennsporteinsatz.* 3. Tagung „Gasfahrzeuge". Berlin 2008.

[95]   BERNER H.-J., BARGENDE M.: *Erdgas als alternativer Kraftstoff – Ein Überblick.* Tagung Innovative Fahrzeugantriebe, Dresden, 2000.

[96]   FERRARI A.: *CFD Analysis of a turbocharged CNG small Displacement Engine.* Diplomarbeit IVK – Universität Stuttgart and Politecnico di Torino, 2004.

[97]   SOSIO M.: *The geometrical compression ratio and crevice behavior evaluation in a 3D-CFD turbocharged CNG Engine.* Diplomarbeit IVK – Universität Stuttgart and Politecnico di Milano, 2009.

[98]   RA Y., KONG S.C., REITZ R.D., RUTLAND C.J., HAN Z.: *Multidimensional Modeling of Transient Gas Jet Injection Using Coarse Computational Grids.* SAE Paper 2005-01-0208.